Climate, Society and the Environment

Climate, Society and the Environment

Rakesh Kumar

RANDOM PUBLICATIONS
NEW DELHI (INDIA)

Climate, Society and the Environment

ISBN 978-93-5111-501-4

Published in 2015 in India by

RANDOM PUBLICATIONS
4376-A/4B, Gali Murari Lal, Ansari Road
New Delhi-110 002
Phone : +9111-43580356, 011-23289044, 011-43142548
e-mail: sales@randompublications.com,
info@randompublications.com, randomexports@gmail.com

Reprinted 2023

Type Setting by : Friends Media, Delhi-110089
Digitally Printed at : Replika Press Pvt. Ltd.

Preface

Climate change has brought about severe and possibly permanent alterations to our planets' geological, biological and ecological systems. There is new and stronger evidence that most of the warming observed over the last 50 years is attributable to human activities. These changes have led to the emergence of large-scale environmental hazards to human health, such as ozone depletion, loss of biodiversity, stresses to food-producing systems and the global spread of infectious diseases.

To date a neglected aspect of the climate change debate, much less research has been conducted on the impacts of climate change on health, food supply, economic growth, migration, security, societal change, and public goods, such as drinking water, than on the geophysical changes related to global warming. Human impacts can be both negative and positive. Climatic changes in Siberia, for instance, are expected to improve food production and local economic activity, at least in the short to medium term. Numerous studies suggest, however, that the current and future impacts of climate change on human society are and will continue to be overwhelmingly negative.

Climate change is about the relationship of society with nature and economy. It is also about the 'nature' of human society, our wants, needs and greed. But too little is said about this connection between science and society. This book joins the dots, to bring a rich understanding of how society can 'fix' this existential challenge by 'fixing' itself first. It represents a valuable source of knowledge covering the science of climate change and vital impacts on society not only at the local level but globally. The value of this book lies in the fact that it informs the public on why action by human society in dealing with climate change is critical and urgent.

Author

Contents

1

Environment and Society

It is only in recent decades that the concepts associated with sustainability have come into more common use. Environmental concerns began to develop in the 1960s, and were particularly driven by Rachel Carson's book *Silent Spring* and the publicity surrounding it. Like other popular and scientific studies at the time, it focused on the environmental harm caused by one economic sector, in this case agriculture. In the 1970s, the Club of Rome identified the problems that societies would face when environmental resources were overused, depleted or harmed, and pointed towards the need for different types of policies to maintain and generate economic growth. In the 1980s, the World Commission on Environment and Development, chaired by Gro Harlem Brundtland, published *Our Common Future,* the first serious attempt to link poverty to natural resource management and the state of the environment. Sustainable development was defined as 'meeting the needs of the present without compromising the ability of future generations to meet their own needs.' The concept implied both limits to growth, and the idea of different patterns of growth, as well as introducing questions of intergenerational justice.

In 1992, the UN Conference on Environment and Development was held in Rio de Janeiro, taking forward many themes prefigured at the UN Conference on the Human Environment held in Stockholm in 1972. The principles of sustainable forms of development that encouraged minimizing harm to the environment and human health were agreed. However, progress has not been good, as Agenda 21 was not a binding treaty on national governments, and all are free to choose whether they adopt or ignore such principles. The Rio Summit was followed by some international successes, including the signing of the Convention on Biodiversity in 1995, the Kyoto Protocol in 1998 and the Stockholm Convention on Persistent Organic Pollutants in 2001. The ten years after the Rio World Summit on Sustainable Development was then held in Johannesburg in 2002, again raising the profile of sustainability, but also failing to tie governments to clear actions and timetables.

Over time, the concept of sustainability has grown from an initial focus on environmental aspects to include first economic and then broader social and political dimensions.

- *Environmental or ecological* - the core concerns are to reduce negative environmental and health externalities, to enhance and use local ecosystem resources, and preserve biodiversity. More recent concerns include broader recognition of the potential for positive environmental externalities from some economic sectors (including carbon capture in soils and flood protection).
- *Economic* - economic perspectives recognize that many environmental services are not priced by markets and that, because of this, it may be economically rational to use the environment in unsustainable ways and to undersupply environmental public goods. In response to this, some seek to assign value to environmental goods and services, and also to include a longer time frame in economic analysis. They also highlight subsidies that promote the depletion of resources or unfair competition with other production systems. *Social and political* - there are many concerns about the equity of technological change. At the local level, sustainability is associated with participation, group action and promotion of local institutions and culture . At the higher level, the concern is for enabling policies that target preservation of nature and its vital goods and services. Many believe that liberal democracies are more likely to give rise to such policies than are autocracies, as part of generally better governance , but the empirical evidence for this is ambiguous. Partly because of this some argue that the liberal democratic state needs to be transcended by adding in representation of other species, other generations and other nations and by enhancing the potential for open deliberation about the issues, to bring together

the knowledge that different groups and communities have and to reduce the corrosive impact of narrow self-interest.

Social Perspective on Environment and Society

An important feature of this Handbook centres on how social organization constrains humans' relationships with nature, but also how social organizations are shaped by nature. Perhaps the most distinctive feature of such an approach is that it rejects the notion that any form of social organization or structured human action is ideal or given by nature. While much human action is constrained by social structures (e.g. market behaviour), it is assumed that those structures are socially constructed and subject to change. This stance implies that human behaviour in relation to nature can be redirected if social structures change. Furthermore, changes in nature may force changes in social structure which in turn lead to changes in human behaviour.

Social scientists have long striven to develop an understanding of the relationship between the natural environment and society, but until the 1970s treatment by sociologists of this relationship remained more implicit than explicit. At this time, sociologists began to consider the nature-society nexus, and contemporary environmental sociology became a reaction to growing social activism for environmental protection. This activism reflected discontent with the dominant pro-technology and pro-growth economic policies following World War II. During the Cold War era, these policies might have tended to be either more market- or state-centred, but regardless of ideological orientation economic growth driven by technological innovation was the overarching approach to economic development. This dominant worldview held that human domination of nature was unproblematic from a practical standpoint and was morally justified as well. But this point of view came to be challenged on both practical and moral grounds.

From a practical standpoint, environmental deterioration became visible to the untrained eye. Air and water pollution became public issues of great concern. Although the scientific community had been the foundation of technological development, critics of various technologies began to emerge from within it as well. Perhaps the most celebrated scientist to mount a sustained critique of the environmental impacts of technology was Rachel Carson. Many observers claim that the publication of her book, *Silent Spring*

, marked the rise of contemporary environmentalism in the USA but there is clear evidence that concern about environmental destruction had already been stirring throughout the industrial world. The rise of the environmental movement in the USA, for example, led to the enactment of a variety of unprecedented environmental legislation.

Sociologists were somewhat taken by surprise by the environmental movement, and struggled to understand it. Its substantive focus as well as the composition of its adherents appeared to be somewhat different from the other social movements of the day. The movement's adherents were initially thought to be more middle class and perhaps more mainstream than the anti-war and civil rights activists of the time. Substantively, the movement seemed to be charting a new course that was not rooted in the dominant socialist or capitalist ideologies. For this reason some sociologists began to suspect that environmentalists were advocating an entirely new paradigm - one that politically was neither left nor right, but entirely different. For this reason some initial thinking by sociologists was that an entirely new theoretical underpinning would need to be formulated.

Initially, existing social theories were largely rejected on the assumption that they had been deficient in considering the active part played by the natural environment in societal development and had considered the impact of society on nature as inconsequential. Without a clear theory to guide the development of an alternative sociology of the environment, early efforts moved in a variety of directions that steered environmental sociology away from established theories of society.

Environmental sociologists initially criticized existing social theories for their hubris in assuming that humans through science and technology could dominate nature without significant impacts on the natural world or society. This paradigm was labelled 'human exemptionalism'. It was immediately clear that any sociology of the environment would need to focus on the relationship between that natural environment and society. A more careful treatment of this issue would challenge many assumptions in sociology. For example, sociology had assumed that all social structures could be explained by human agency. From this point of view, the physical and biological worlds were passive objects in the human construction of the social world. But environmentalists' concerns about the destruction of nature and its consequences for society led to a reconsideration of how nature shapes society. Some claimed that what was distinctive about environmental sociology was its emphasis on the mutual constitution of nature and society. From this perspective, some sort of unidirectional and exclusively human construction of the life world is impossible.

So, what shapes the relationship between society and the environment? Some early attempts to apply sociological theory to the understanding of nature-society relationships drew on Marxist political economy. Political economists focused on the nature of the capitalist organization of production and how the functional demands of this system defined the use of nature. Some of the early thought in this area emphasized how capitalism's requirement for the continuous expansion of production into new areas would inevitably lead to the destruction of nature. More recently, there has been greater emphasis on how capitalism is constrained by the biological and physical limits imposed by the natural world.

The implications of the dominant system of market capitalism for nature-society relationships are a point of considerable contention in sociology. Some would argue that the capitalist economy is fundamentally destructive of the environment and for this reason is unsustainable in the long run. From this point of view, environmental destruction is the 'Achilles heel' of capitalism. This approach is deeply suspicious of claims that science and technology can always produce adequate substitutes for depleted natural resources. Recently, a decidedly more optimistic theory of ecological modernization has come into play. From this point of view, environmental destruction reflects a lack of investment in modern technologies and this deficit can be remedied with state policies that prohibit production practices wasteful or destructive of the environment. Ecological modernization is not just about technology, though. It is as much about bringing ecological considerations into market decision making through appropriate pricing of environmental services. In this theory the state plays a prominent role, with little real significance attached to abstractions like the 'free' market. The state constrains markets through policies that establish incentives to channel market behaviour in environmentally sound directions.

These opposing viewpoints on the environmental impacts of market economies point to the distinctiveness of this approach to understanding nature-society relations. Regardless of their theoretical orientation, sociologists consider organizational forms to be social constructs that are subject to change. This assumption implies that human behaviour is not inherent or given, but moulded by the social structures in place at any time in history. Thus, sociologists emphasize the distinctiveness of processes of societal rationalization, or the elaboration of a historically specific logic that structures the interaction between nature and society. Any particular rationalization is not 'natural' but has a distinctive form that constrains options for human interactions with nature.

Since sociologists assume that social organization does not take some sort of 'ideal' form, the organization of human interactions with nature is a subject of particular interest to sociologists. Given an infinite number of possible forms of organization, why are similar forms of organization widely dispersed across a wide range of social and natural environments? This question has become especially salient with the emergence of the processes of globalization. Economic, environmental and social organization displays some striking similarities in far-flung parts of the world. This organizational isomorphism is of growing interest to sociologists. But just as interesting to sociologists are some of the distinctive ways that these organizations are refashioned by local interests and the local natural resource base.

Environmental Assets and Externalities

Many economic sectors directly affect many of the very assets on which they rely for success. Economic systems at all levels rely on the value of services flowing from the total stock of assets that they influence and control, and five types of asset, natural, social, human, physical and financial capital, are now recognized as being important. There are, though, some advantages and misgivings with the use of the term capital. On the one hand, capital implies an asset, and assets should be cared for, protected and accumulated over long and intergenerational periods. On the other, capital can imply easy measurability and transferability. Because the value of something can be assigned a monetary value, then it can appear not to matter if it is lost, as the required money could simply be allocated to purchase another asset, or to transfer it from elsewhere. But nature and its wider values is not so easily replaceable as a commodity.

Nonetheless, as terms, natural, social and human capital have become widespread in helping to shape concepts around basic questions about the potential sustainability of natural and human systems. The five capitals have been defined in the following ways:

1 *Natural capital* produces environmental goods and services, and is the source of food (both farmed and harvested or caught from the wild),

wood and fibre; water supply and regulation; treatment, assimilation and decomposition of wastes; nutrient cycling and fixation; soil formation; biological control of pests; climate regulation; wildlife habitats; storm protection and flood control; carbon sequestration; pollination; and recreation and leisure.

2 *Social capital* yields a flow of mutually beneficial collective action, contributing to the cohe-siveness of people in their societies. The social assets comprising social capital include norms, values and attitudes that predispose people to cooperate; relations of trust, reciprocity and obligations; and common rules and sanctions mutually agreed or handed down. These are connected and structured in networks and groups.

3 *Human capital* is the total capability residing in individuals, based on their stock of knowledge skills, health and nutrition. It is enhanced by access to services that provide these, such as schools, medical services and adult training. People's productivity is increased by their capacity to interact with productive technologies and with other people. Leadership and organizational skills are particularly important in making other resources more valuable.

4 *Physical capital* is the store of human-made material resources, and comprises buildings, such as housing and factories, market infrastructure, irrigation works, roads and bridges, tools and tractors, communications, and energy and transportation systems, that make labour more productive.

5 *Financial capital* is more of an accounting concept, as it serves in a facilitating role rather than as a source of productivity in and of itself. It represents accumulated claims on goods and services, built up through financial systems that gather savings and issue credit, such as pensions, remittances, welfare payments, grants and subsidies.

As economic systems shape the very assets on which they rely for inputs, there are feedback loops from outcomes to inputs. For instance, some economists emphasize the way that markets respond to resource scarcity is by pushing up prices, encouraging substitution and searching for technical change. However, such market feedbacks cannot work properly if environmental assets come for free. Thus, while sustainable systems will have a positive effect on natural, social and human capital, unsustainable ones feed back to deplete these assets, leaving fewer for future generations. For example, an agricultural system that erodes soil whilst producing food externalizes costs that others must bear. But one that sequesters carbon in soils through organic matter accumulation helps to mediate climate change. Similarly, a diverse system that enhances on-farm wildlife for pest control contributes to wider stocks of biodiversity, whilst simplified modernized systems that eliminate wildlife do not. Agricultural systems that offer labour-absorption opportunities, through resource improvements or value-added activities, can boost local economies and help to reverse rural-to-urban migration patterns.

Any activities that lead to improvements in these renewable capital assets thus make a contribution towards sustainability. However, the idea of sustainability does not suggest that all assets are improved at the same time. One system that contributes more to these capital assets than another can be said to be more sustainable, but there may still be trade-offs with one asset increasing as another falls, though some environmental assets are essentially irreplaceable and vital, so they cannot be substituted. Many economic systems are, therefore, fundamentally multifunctional. They jointly produce many environmental goods and services. Clearly, a key policy challenge, for both industrialized and developing countries, is to find ways to maintain and enhance economic productivity. But a key question is: can this be done whilst seeking both to improve the positive side-effects and to eliminate the negative ones? It will not be easy, as modern patterns of development have tended to ignore the considerable external costs of harm to the environment.

Valuing the Environment

The idea that the environment and the services it provides can be valued strikes some as antithetical to the intrinsic values of environmental resources and the role that these resources play in society, history and culture. How can we possibly assign an economic or monetary value, it might be asked, to unique biodiversity such as the bald eagle or the snow leopard, to views of the Alps or the Rocky Mountains, or to water resources that are essential to life and that many societies consider to be an inherent human right? If economic/monetary values of these and similar resources can be estimated, how can they possibly be accurate if underlying conditions of scarcity change, as they inevitably will, leading to changes in associated scarcity values? And, if economic/monetary values are assigned to resources, whatever those values may be, does this valuation in and of itself inevitably lead to political trade-offs that may degrade those resources in the interests of economic development or other goals?

For these and many other reasons, the valuation of environmental resources is often fraught with contention, both conceptually and certainly in practice, where many empirical estimation and measurement issues arise. Yet, as mentioned above, the treatment of environmental assets as natural capital and associated exercises in measurement, valuation and evaluation are increasingly common in both academic analysis and policymaking. This is for several reasons. First, without such valuations, society has done a remarkably poor job in managing its stewardship of environmental resources; surely, any mechanism that can help improve on society's past dubious record in environmental policy is an advance. Second, since at least the 1960s and 1970s, the environmental impacts of economic development and human interventions in the landscape have been central to policy debates as society has increasingly been concerned with both the direct effects and opportunity costs of those interventions - e.g. what is lost when development proceeds. Third, in the two decades since the publication of the Brundtland Report , issues of sustainability have achieved much higher prominence in public debate in many countries, highlighting the needs of future generations in decisions made today about resource use. This has increased interest in how to trade off current versus future demands on the environment and how to deal with associated intergenerational equity concerns, which, in turn, has increased interest in mechanisms, like economic valuation, that permit these intertemporal comparisons.

In addition to these general factors stimulating interest in environmental valuation, efforts at economic and monetary valuation of the environment have flourished over the past several decades because they address several additional specific needs that are increasingly evident in environmental policymaking. First, the importance of the divergence between social valuation of resources and their incomplete (or non-existent) valuation in the market is increasingly apparent. How can we begin to address the problem of global warming, for example, if the externalities of industrial pollution are so poorly measured and understood, and consequently devalued in the policy arena, compared with the measurable jobs and income that are created? Second, as the human population expands and many formerly abundant resources are increasingly scarce - clean water and clean air, wilderness, open space, even silence - accounting for, and valuing, the public good dimensions of these resources has become increasingly important in prioritizing their survival in policy debates. How else, outside of moral suasion, will the scarcity value of public goods be understood and taken into account? Third, as the demand for economic valuation has expanded since the 1960s and 1970s, specific valuation methods and estimation procedures have also improved significantly, permitting a more accurate - though still frequently problematic - estimation of economic and monetary values of environmental resources and associated services.

An additional factor has to do with the response to policymaking itself. The limitations of 'command and control' and 'fences and fines' approaches to environmental policymaking have become increasingly evident, both in industrialized countries, where the institutions are often in place to deal effectively with at least some environmental problems, and certainly in developing countries, where such institutions are often nonexistent, irrelevant or functionally powerless. Yet, 'command and control' policy and regulatory approaches often generate responses by private decision-makers that are, at best, socially inefficient and wasteful of resources, and, at worst, stimulate rent-seeking behaviour and strategic decision-making that yield perverse outcomes. Is it not preferable to develop policies and regulatory frameworks that are compatible with private incentives and that, in fact, employ these incentives and knowledge of human behaviour in innovative ways to lead to socially desired outcomes? Much of the recent interest in environmental valuation has been concerned with precisely these questions, specifically, the development of incentive-compatible policies and regulatory approaches that yield desired outcomes in ways that may be less costly and more socially efficient. Hence, the interest in tradable emissions permits, carbon-trading schemes, the pricing of heretofore free water resources, valuation, compensatory and payment transfer mechanisms for environmental services, and other such innovations.

Although alternative typologies exist, one common framework for organizing our thinking about resource valuation distinguishes four types of ecosystem values : (1) *direct use values,* due to the direct utilization of resources and ecosystem services; (2) *indirect use values,* attributable to the externalities of ecosystem services; (3) *option value,* due to preserving the option for future use of the resource (also directly addressing sustainability criteria); and (4) *non-use values,* which are attributable to a variety of intrinsic ecosystem characteristics. This nomenclature aside, perhaps inevitably, much of the attention in environmental valuation has focused on specific methodologies and analytical approaches to assigning economic and monetary values to resources, especially those resources that have typically been outside the formal market.

The former approaches ask consumers and other private agents to assign resource values and rankings directly; these approaches include 'contingent valuation' methodologies in which people are asked for their 'willingness-to-pay' to pay for environmental benefits, for example. The latter approaches indirectly elicit consumer valuations through methods such as the 'travel cost' approach and 'hedonic pricing', which estimate resource values through statistical analysis of factors underlying human behaviour and the preferences (e.g. values) that are thus revealed. All of these methods have acknowledged strengths and deficiencies. Yet, they have achieved wide acceptance because they continue to be at least partially successful in giving policy analysts and policymakers useful mechanisms and standards for achieving a better understanding of the values of environmental resources, thus enabling them to make better decisions regarding resource management, including the conservation and preservation of environmental resources in the face of competing uses.

Consumption Treadmill

Since the World Commission on Environment and Development began deliberating on the links between environment and economy, there have been at least a couple of hundred further definitions of sustainability, and the term has now entered our common language. But where are we now with this sustainability idea? Does it offer some new hope for the world, or has it just hidden a much greater problem? The biggest challenge to sustainable development is now the consumption treadmill. The figures are worrying. People in North America now consume 430 litres of water per day; in developing countries, 23% have no water. In North America, there are 75 motor vehicles per 100 people, in Japan 57, in Europe 24, and in China, India and Africa just six to nine. Worldwide, some 400,000 hectares of cropland are paved per year for roads and parking lots. The world motor-vehicle fleet grows alarmingly, as the nearly wealthy look to other parts of our global community for guidance as to what to buy. By almost every measure of resource consumption or proxy for waste production, the USA and Europe lead the way. And what model is being held up as the one to aspire to? There are now few people in the world who do not now aspire to the same levels of consumption as North America, which is, after all, presented as the pinnacle of economic achievement.

Table 1. Indicators of consumption from different countries and regions of the world (data from 2004-05)

	USA	*Europe*	*China*	*India*	*Asia*	*Africa*	*Latin Central America*	*World*
Passenger cars per 100 people	75	24	7	6	17	9	6	9
Annual petrol and diesel consumption (litres per person)	1624	286	33	9	47	36	169	174
Annual energy consumption per person (kg oil equivalent)	8520	3546	896	515	892	580	1190	1640
Annnual carbon dioxide emissions (tonnes per person)	20.3	8-12	2.7	0.99	<1	<1	<1	3.85
Annual paper and board consumption (kg per person)	308	125	34	4	29	4	38	52
Annual meat consumption (kg per person)	125	74	52	5	28	13	58	40
Daily water consumption (cubic metres per person)	4.6	1.59	1.35	1.74	1.72	0.47	1.47	1.73
Population	293	730	1306	1080	3667	887	518	6500
Children born per woman	2.08	1.56	1.72	2.78	3.1	4.82	2.75	2.55

This consumer boom is already happening. The new consumers have already entered the global economy, and are aspiring to have lifestyles currently enjoyed by the richest. A number of formerly poor countries are seeing the growing influence of affluence, as the middle classes of China, India, Indonesia, Pakistan, Philippines, South Korea, Thailand, Argentina, Brazil, Colombia and Mexico engage in greater conspicuous consumption. The side-effects are already being felt - the average car in Bangkok spends 44 days a year stuck in traffic. But there is still a long way to go. The car fleet of the whole of India is still smaller than that of Chicago, and that of China is half the number of cars in greater Los Angeles. At the same time as a consumer boom is occurring among newly affluent urban elites, poor people in such countries as India and China lack access to the basics such as clean water and health care.

This is now the concern: the idea of sustainable economic development seems to imply that the world can be improved, or even saved, by bringing everyone up to the same levels of consumption as those in the industrialized countries. We can, it is said, grow out of many kinds of economic trouble. This cannot be done, as we would need six worlds at European and eight to nine at North American levels and patterns of consumption. How much, we might wonder, would be enough?

The currently dominant idea about the inevitable benefits of progress would appear to be a modern invention. Indigenous peoples do not believe that their current community is any better than those in the past. To them, past and future are the same as current time. Their ancestors, and those of animals too, constantly remind them to be humble as they move about their landscapes. But the myth of progress permits the losses of both species and special places, as it is believed that losses can be offset by doing something else that is better. The myth permits a belief in technological fixes, which are indeed effective in many ways, but rarely seem to make everyone happier, even if some of them contribute to human longevity and reduce suffering. Environmental problems are, after all, human problems. New technologies will make improvements, but possibly not fast enough to save us. They also bring some new risks, possibly rendering society more vulnerable. To come soon will be fabulous electronic memory, a genomics revolution, renewable energy, and human brains augmented by computers, though as Rees puts it, 'a super-intelligent machine could be the last invention humans ever make.' Rees recounts the 1937 efforts by the US National Academy of Sciences to predict breakthroughs for the rest of the last century. They made a good stab at agriculture, rubber and oil, but completely missed nuclear energy, antibiotics, jet aircraft, space travel and computers.

It is now clear from a variety of studies of people in the USA and Europe that people were happier in the 1950s compared with today. We can only guess more about earlier times, as the data do not exist in comparable form. But it does seem that our programmed happiness is about striving for, not actually increasing, happiness. One reason is that we compare our consumption with others around us, and we do not necessarily feel better off or happier if others' consumption is also increasing. There is always a nagging gap between present levels of contentment and how it could be. We believe we will be happier in the future, but seldom are. We also are constantly worrying about how future life events affect our happiness. As Bell has pointed out, we could work four hours per day, or just for about half a year, if we consumed at 1940s levels, yet be equally happy. But would anyone choose this option if they could?

Population and the Environment

Population will continue to grow in many countries at least until mid-century, posing considerable problems in relation to providing for basic needs and dealing with environmental damage in some. Yet population is already declining is some rich countries, and others' population can be expected to stabilize then to decline, as the age structure of the population shifts and social practices change. A psychological problem yet to be faced is the consequence of coming population decline. Thomas Malthus argued that human population growth would always outstrip resources. 'Population, when unchecked,' he said, 'increases in a geometrical ratio. Subsistence increases only in an arithmetric ratio. A slight acquaintance with numbers will shew the immensity of the first power in comparison with the second'. Since then, most policies and practice regarding natural resources and food have been shaped by concerns about our growing numbers. Humans are, after all, an extraordinarily successful species. When agriculture emerged, some 10,000 years ago, there were probably five million people worldwide. To the mid-19th century, world population then doubled eight times. Since then it has doubled four more times, and will continue to grow to probably eight and a half billion people by the middle of the 21st century. It will then stabilize for a while, and subsequently fall. Not because of wars, climate change or infectious diseases (though they may contribute to greater declines), but because of changing fertility patterns. More choices about contraception and decreasing poverty reduces the need to have so many children, and changing lifestyles among the rich delay child-bearing ages. When one generation produces fewer daughters, and fewer daughters are produced by them, then the replacement rate soon falls below the 2.1 needed to maintain population stability.

Today, the average woman in industrialized countries has fewer than 1.6 children, in the least developed countries 5 children, and in the other developing countries 2.6. The lowest fertility rates are now in southern Europe, at 1.1 children per woman. In the mid-1970s, the average Bangladeshi woman had six children; today she has about three; in Iran, fertility has fallen from more than five children in the late 1980s to just over two today. The worldwide annual gain is still 76 million people , but this is expected to fall to zero by 2050 as the number of children falls from today's average of 2.55 to 2.0. Life expectancy at birth was 47 years in 1950-55, rose to 65 years by 2000-05, and will rise again to 75 years worldwide by 2045-50. By then, the number of people over 60 will have tripled to 1.9 billion, and the number over 80 will have risen from today's 86 million to 395 million. Of course, these changes will not be evenly spread. Some countries are predicted to triple their numbers by 2050: these include Afganistan, Burkina Faso, Burundi, Chad, Congo, DR Congo, DR Timor-Leste, Guinea-Bissau, Liberia, Mali, Niger and Uganda. But the populations of 51 countries will fall, including Germany, Italy, Japan and most of the former USSR.

What will happen after this peak, less than two generations away from us now? The United Nations has made population predictions for the next 300 years, and uncertain though these must be, the medium fertility estimates suggest at least a levelling of world population for 250 more years at 8.5 to 9 billion. At low fertility (at the kind of levels we are already seeing today - after all, 93 out of 222 countries already have fewer than 2.1 children per women, and 37 have less than 1.5), world population declines to 5.5 billion by the end of this century, to 3.9 billion by 2150, and down to 2.3 billion by 2300. Which track we end up on depends entirely on early changes in fertility. Demographers cannot, of course, agree on the probability of stability or decline. But any kind of fall will bring huge changes. In 2000, people on average retired two weeks before mean life expectancy (at 65 years); by 2300, people will retire more than 30 years short of life expectancy (unless age of retirement changes), when on average women will live to 97 and men to 95 years. This does not take account of potentially revolutionary changes to human longevity that new medical technologies might bring.

Caldwell says that 'the low scenario is by no means implausible', and that the low projections 'would probably portend to many the fear of human extinction'. Governments would try to raise fertility levels, but it could be very difficult to achieve, as people do not always do the bidding of their governments. What, then, will happen to all those settlements we do not need? What of the fields and farms that become surplus to requirements? What of the wild animals - will we see their return to places where they had long since been eliminated (not the extinct species, of course, as they are gone forever)? Or might the vision be quite different - of spreading urban wastelands, of forgotten linkages to nature, of the nightmare of decivilization.

Dualism, Separation and Connections

In recent years, with growing concerns for sus-tainability, the environment and biodiversity, many different typologies have been developed to categorize shades of deep to shallow green thinking. Arne Naess sees shallow ecology, for example, as an approach centred on efficiency of resource use, whereas deep ecology transcends conservation in favour of biocentric values. Other typologies include Donald Worster's imperial and arcadian ecology and the resource and holistic schools of conservation. For some, there is an even more fundamental schism -whether nature exists independently of us, or whether it is characterised as post-modern or as part of a post-modern condition. Nature to scientific ecologists exists. To some post-modernist perspectives, though, it is mostly a cultural construction. The truth is, surely, that nature does exist, but that we socially construct its meaning to us. Such meanings and values change over time, and between different groups of people.

There are many dangers in the persistent dualism that separates humans from nature. It appears to suggest that we can be objective and independent observers - rather than part of the system and inevitably bound up in it. Everything we know about the world we know because we interact with it, or it with us. Thus, if each of our views is unique, we should listen to the accounts of others and observe carefully their actions. Another problem is that nature is seen as having boundaries -the edges of parks or protected areas. At the landscape level, this creates difficulties, as the whole is always more important than each part, and diversity is an important outcome.

This can lead to the idea of enclaves - social enclaves such as reservations, barrios or Chinatowns, and natural enclaves like national parks, wildernesses, sites of special scientific interest, protected areas or zoos. Enclave thinking can lead us away from accepting the connectivity of nature and people, though it has the advantage of creating niches for specialization.

One consequence is that biodiversity and conservation can be considered to be in one place, and productive agricultural activities in another. It is no longer acceptable to cause damage in some natural landscapes, provided we leave some areas protected. Enclaves also act as a sop to those with a conscience - the wider destruction can be justified if we fashion a small space for natural history to persist.

By continuing to separate humans and nature, the dualism also appears to suggest that technologies can always intervene to reverse damage caused by this very dualism. The greater vision, and the more difficult to define, involves looking at the whole, and seeking ways to redesign it. Cartesian dualism that puts humans outside nature remains a strange concept to many human cultures. It is only modernist thinking that has separated humans from nature in the first place, putting us up as distant controllers. Most peoples do not externalize nature in this way. From the Ashéninha of Peru to the forest dwellers of former Zaire, people see themselves as just one part of a larger whole, as do many people who adhere to major modern religions - even Christians who are often accused of treating nature as something to be plundered. Their relationships with nature are holistic, based on 'both/with' rather than 'either/or'. Recent research on the biophilia hypothesis of E. O. Wilson is indicating that natural or green places are good for mental health, irrespective of social context.

The idea of the wilderness struck a chord in the mid-19th century, with the influential writers Henry David Thoreau and John Muir setting out a new philosophy for our relations with nature. This grew out of a recognition of the value of wildlands for people's well-being. Without them, we are nothing; with them, we have life. Thoreau famously said in 1851, 'in wildness is the preservation of the world'. Muir in turn indicated that: 'wildness is a necessity; and mountain parks and reservations are useful not only as fountains of timber and irrigating rivers, but as fountains of life.' But as Roderick Nash, Max Oelschlaeger, Simon Schama and many other recent commentators have pointed out, these concerns for wilderness represented much more than a defence of unencroached lands. It involved the construction of a deeper idea, which proved to be hugely successful in reawakening in North American and European consciences the fundamental value of nature.

Debates have since raged over whether 'discovered' landscapes were 'virgin' lands or 'widowed' ones, left behind after the death of indigenous peoples. Did wildernesses exist, or did we create them? Donald Worster, environmental historian, points out for North America that 'neither adjective will quite do, for the continent was far too big and diverse to be so simply gendered and personalised'. In other words, just because they constructed this idea does not mean to say it was an error. Nonetheless, they were wrong to imply that the wildernesses in, say, Yosemite were untouched by human hand, as these landscapes and habitats had been deliberately constructed by Ahwahneechee and other native Americans and their management practices to enhance valued fauna and flora.

Henry David Thoreau developed his idea of people and their cultures as being intricately embedded in nature as a fundamental critique of mechanical ideas that had separated nature from its observers. His was an organic view of the connections between people and nature. In his *Natural History,* Thoreau celebrates learning by 'direct intercourse and sympathy' and advocates a scientific wisdom that arises from local knowledge accumulated from experience combined with the science of induction and deduction. But he still invokes the core idea of wilderness untouched by humans - even though his Massachusetts had been colonized just two centuries earlier and had a long history of 'taming' both nature and local native Americans.

The question, 'is a landscape wild, or is it managed', are perhaps the wrong ones to ask, as it encourages unnecessary and lengthy argument. What is more important is the notion of human intervention in a nature of which we are part. Sometimes such intervention means doing nothing at all, so leaving a whole landscape in a 'wild' state, or perhaps it means just protecting the last remaining tree in an urban neighbourhood or hedgerow on a field boundary. Preferably, intervention should mean sensitive management, with a light touch on the landscape. Or it may mean heavy reshaping of the land, for the good or the bad.

So it does not matter whether untouched and pristine wildernesses actually exist. Nature exists without us; and with us is shaped and reshaped. Most of what exists today does so because it has been influenced explicitly or implicitly by the hands of humans, mainly because our reach has spread as our numbers have grown, and as the effects of our consumption patterns have compounded the effect. But there are still places that seem truly wild, and these exist at very different scales and touch us in different ways. Some are on a continental scale, such as the Antarctic. Others are entirely local, a woodland amidst farmed fields, a saltmarsh along an estuary, a mysterious urban garden, all touched with private and special meanings.

In all of these situations, we are a part, connected, and so affecting nature and land, and being affected by it. This is a fundamentally different position to one which suggests that wilderness is untouched, pristine, and so somehow better because it is separated from humans - who, irony of ironies, promptly want to go there in large numbers precisely because it appears separate. But an historical understanding of what has happened to produce the landscape or nature we see before us matters enormously when we use an idea to form a vision that clashes with the truth. An idea that this place is wild, and so these local people should be removed. Another idea that this place is ripe for development, and so a group of people should be dispossessed. The term wilderness has come to mean many things, usually implying an absence of people and presence of wild animals, but also containing something to do with the feelings and emotions provoked in people. Roderick Nash takes a particularly Eurocentric perspective in saying, 'any place in which a person feels stripped of guidance, lost and perplexed may be called a wilderness', though this definition may also be true of some harsh urban landscapes. The important thing is not defining what it really is, but what we think it is, and then telling stories about it.

Sociology and the Environment

The classical approaches to understanding the structure of society shared two basic features. One was the ambition to provide ways of conceptualizing the large-scale structural features of whole societies, and to situate them in the context of long-term historical change and in relation to the alternative social forms and historical tendencies in the rest of the world. The other was the insistence that human social and historical life was a distinct order of reality in its own right, not to be explained away in terms of the biological sciences of the day: industrial development, social inequality, crime, suicide rates, gender divisions and the like were to be understood in terms of social and cultural causes, not racial inheritance, genetic endowment or physiological constitution. This second feature was the basis for a process of 'separate development', through which the life and social sciences proceeded in ignorance of one another: 'nature' and 'culture' were distinct and contrasting realms, knowing and needing to know nothing of one another.

A common feature of the classics was their insistence on human social and cultural life as an order of reality in its own right, irreducible to the biological realm. Through most of the 20th century, this was taken to be an unquestioned assumption: social processes were to be explained in terms of social causes. This resistance to biological explanation was strongly reinforced by widespread revulsion at the consequences of Nazi doctrines of racial superiority, and the racist underpinning of much European imperial domination of non-Western peoples. With the rise of new social movements from the 1960s onwards, challenging established inequalities and social exclusions based on gender difference and sexual orientation, the terms 'nature' and 'natural' came to be viewed with suspicion. Sociologists sympathetic to the struggle for women's emancipation or gay rights critically exposed the way dominant ideologies justified oppression in the name of a distinction between what was 'natural' and what was 'unnatural', and therefore pathological. In this way, the strong links between sociology and progressive social movements reinforced the assumption already built into the main sociological traditions that biology, the 'natural', should be held at arms length and viewed with suspicion. It became a standard procedure for sociologists, and especially those who identified with a critical stance towards established society, to call into question all authoritative claims to knowledge of 'nature' or 'reality'.

Then, from the late 1970s onwards, developments in linguistics and cultural theory became very influential, and approaches which (following Weber and Simmel, among the classics) focused on symbolic meaning and the role of language in shaping our experience of the world flourished. Questions about the material reality of nature and our relation to it now became excluded as a matter of methodological principle: all experience of the world is to be understood as mediated by language and culture. But there is no way anyone can stand outside the available language and culture to see reality in itself: we are left with the task of characterizing the role played in social life by various different and often conflicting linguistic and cultural 'constructions' of reality. It is important to remain neutral and agnostic about which, if any, of these 'constructions' is true. Critical sociology can aid emancipatory social struggles by exposing the 'constructed' character of the prevailing oppressive accounts of what is 'natural', thus 'deconstructing' them and challenging their authoritative hold over peoples' lives. These are the core insights of the approaches called 'constructionist'.

The sensitivity of sociology to the social and cultural movements and issues in the wider world outside the academy now presented it with a deep challenge: from the early 1960s the progressive social movements with which many sociologists had become identified also included a burgeoning radical environmental movement whose intellectual leaders (often dissident natural scientists) raised public alarm about the growing threat posed by our affluent, growth-oriented throwaway society to its own planetary life-support systems. Here was a new and powerful basis for a radical, critical politics, but one which celebrated nature, and claimed authoritative knowledge of the terrible destruction of it unleashed by contemporary society. This was a deep challenge in two ways. First, it was an intellectual challenge. Sociology had established its right to exist as a distinct discipline by a radical separation of the realms of nature and culture, but now faced pressing questions about the consequences of the mutual interconnection, the shared fate, of natural processes and social life. The second challenge was rooted in the normative commitment of critical sociologists and was particularly strongly felt by those who sympathized with such emancipatory movements as anti-racist, gay rights and women's liberation activism but were also drawn to the emergent green politics with its passionate defence of 'nature'.

There emerged two very broad, and to some extent conflicting, ways of addressing the new environmental agenda. One, typically 'constructionist', and deriving from the 'modest' tradition, tended to avoid large-scale theorizing. The great strength of this tradition has been its detailed case studies of particular environmental issues, social movements, campaigns and episodes of conflict. Rather than use the new environmental agenda as an occasion for questioning the basic inherited assumptions of the discipline, this sort of approach has concentrated on treating environmental issues as a new field in which to demonstrate the value of already-established sociological concepts and styles of argument. For instance, this approach has debunked many myths about the environmental movement. It was found that while parts of the movement retain a radical and progressive edge, many had evolved into highly professional lobby organizations seeking insider status in government decision making through moderating their demands. At the same time, many members do little or nothing beyond giving an annual donation, having little or no direct involvement in local politics and living remarkably standard middle-class lives. The key standard concepts turn out to be institutionalization as a consequence of the problem of resource mobilization, which in turn derives from rational choice by individuals to do little, as captured in the Prisoner's Dilemma metaphor and other models of collective action failure.

For sociologists of science, the natural sciences are thoroughly social in character, their conceptual organization, research priorities and methodological procedures all shaped by social interests and cultural values - generally subservient to the dominant group or elite interests. Similarly with technologies - these are designed to serve powerful interests and cannot be fully understood independently of the social practices and relationships which their use either maintains or transforms. This way of understanding scientific and technical innovation has much to offer in the environmental field. It opens up the possibility of analysis of the kinds of pressures, power relationships, forms of regulation, etc., which promote environmentally damaging technologies, and also suggests the sorts of social and economic change which might encourage more benign forms of technical change.

Contructionist approaches have also produced valuable research in the field of environmental social movement mobilization and organization. The key insight which informs their approach is recognition that there is no one-to-one correspondence between the existence of, say, air pollution, or biodiversity loss, on the one hand, and the emergence of a social movement which identifies it as an unacceptable condition and campaigns for change, on the other. A leading constructionist environmental sociologist, John Hannigan , provides an illuminating set of concepts for analysing the social and cultural processes involved in 'constructing' an environmental problem. First, a problem-claim has to be 'assembled': evidence, including scientific evidence, has to be collected and put together in such a way as to show that the state of affairs is significant enough to justify public concern and action. Next, it has to be 'presented': the problem has to be characterized in ways which will attract attention, and provoke the desired public concern. Since the media are now so central to communication to wider publics, this will also involve ways of engaging with the media in such a way as to ensure not only their attention, but also that media representations coincide with the movement's own 'framing' of the issue. The case of Greenpeace's use of dramatic film footage of whaling is a good example. The visual images were irresistible material for the electronic media, and their vivid portrayal of the violent death of great and beautiful creatures had more impact on public conscience than a thousand books. Perhaps, too, the constructionists might argue that the ethical and aesthetic power of these images far outweighed the influence of detailed scientific studies of the population dynamics and risk of extinction of the different whale species.

Finally, Hannigan notes that success on the part of social movements in making their 'problem-claim' is not the end of the matter. In each case, interests will be threatened by the raising of an issue - in the case of whaling, for example, the industry itself, and spin-off processing and retail interests, as well as consumer cultures in certain countries and indigenous people for whom whaling is central to their whole way of life. So, the raising of an issue will generally be met with counter-arguments, and competition for media framing and public acceptance. Hannigan calls this the 'contestation' of movement claims. Social movement theory also has developed concepts for analysing the processes involved in establishing, maintaining and coordinating social movement activity, for studying the culture of such movements and how they shape the identities of individuals who participate in them.

It is increasingly common for constructionists to defend a more limited or 'contextual' constructionism, which does not deny either the reality or the importance of actual environmental change. So, there is a convergence between constructionist and the alternative 'realist' approaches in their underlying philosophy. Even so, the rhetoric of a more radical constructionism is often retained, and a lack of analysis of the crucial ambiguities of concepts such as 'construction' itself can give the impression that an account of the cultural *construction* of an environmental change as an 'issue' somehow also explains the socio-economic *causes* of the change itself. In other words, the constructionist approaches may be true to the 'nature-sceptical' critical traditions, but they do not, in the end, address the need to revise those traditions to reconnect our understanding of society with its material basis in nature.

The four basic types of approach are, first, the 'new environmental paradigm' advanced in the late 1970s, second, 'reflexive modernization', as advocated by Giddens, Beck and others, third, a more recent cluster of approaches referred to as 'ecological modernization', and, finally, a range of approaches deriving from the Marxist, or historical materialist, tradition in various combinations with green, feminist and anti-racist ideas. These latter approaches can be collectively referred to as 'radical political economy'. The pioneers of the first approach were the US sociologists, R.

E. Dunlap and W. R. Catton. In a series of articles from the late 1970s onwards they criticized mainstream sociology for working with a 'human exemptionalist' paradigm: that is, sociologists had tried to understand human societies in abstraction from their interdependence with the rest of nature, as if we were 'exempt' from the laws of nature which apply to all other beings. Instead, they proposed a 'new environmental paradigm' which would locate human societies within the wider web of environmental interactions.

Clearly, their proposal was for an ecology-inspired radical revision of the whole sociological tradition. Very much in line with these original proposals is an influential approach which attempts to measure the scale of materials and energy taken up by and emitted by particular societies at different historical periods. Key concepts in materials and energy flow accounting are the 'metabolism' between societies and nature, and 'colonisation' of nature and natural resources by social processes. This approach offers a means of quantitative measure of the extent of 'ecological modernization' over time in the industrialized countries.

Whereas the focus of the radical political economy analyses is modern capitalism, its expansionary tendencies and political implications, the key concept for these other approaches is 'modernity' and the key process 'modernization'. The shift from 'modern' as an adjective, to the idea of 'modernity' as a way of characterizing a whole society or historical period is associated with a tradition known as 'functionalism'. This approach assumed an evolutionary development in the history of societies toward more complex and functionally differentiated societies. The Western societies represented the highest developmental stage, and models were devised to foster 'development' in the rest of the world, on the assumption that it would follow the model already achieved in the west. This process was 'modernization'. Its outcome would be capitalist and liberal-democratic. 'Modernity' was the state we in the West had already attained, and, by implication, one to which everyone else would, or should, aspire. In this early version the notion included three aspects: modernity was the destiny of the whole world, the West was leading the way, and this was a good thing. Initially influential as a cold-war ideology, this assumed more triumphalist forms with the fall of Soviet and East European state-centralist regimes at the end of the 1980s. This period also marked a revival in the use of the terms 'modernity' and 'modernization' by sociologists, often as a way of avoiding the more politically contentious term 'capitalism'.

Most relevant to our theme have been two theoretical approaches which have linked 'modernity' and 'modernization' with ecological change and environmental social movements: ecological modernization theory and the notion of 'reflexive modernization' associated with Ulrich Beck and Anthony Giddens. Both approaches see 'modernity' as a phase in historical development as well as a type of society, and both subdivide modernity itself into successive developmental phases.

In this respect, reflexive modernization theorists, especially, incorporate some of the themes of post-modernism as characterizing a significant transition within modernity. Beyond this, the two traditions diverge quite radically.

Ecological modernization theory had its origins in the 1970s. Its earliest advocates shared the optimistic evolutionary/developmental perspective of the American functionalist versions. They distinguished an early phase of modernity, a phase of industrial 'construction', in which increased production was won at the cost of increased environmental degradation, from a more recent phase of 'reconstruction'. In this latter phase, industrial production and consumption were increasingly governed by a new, 'ecological rationality'. Scientific and technical innovation was increasingly devoted to adapting the industrial society to environmental constraints.

The idea of reflexive modernization, too, has a two-phase model of development within 'modernity'. Modernity itself is defined (in Giddens's version) as a combination of four distinct institutional dimensions: a liberal democratic state, concerned with surveillance, a military establishment which monopolizes the legitimate use of force, an economic system, characterized by private property and market, and industrial technologies as the mode of appropriation of nature. However, in recent decades, this model of 'simple' modernity is rendered increasingly inappropriate by three interrelated social processes. Globalization, which, for Giddens, is primarily a matter of increased international flows of communication and information, opens up all closed communities and stable traditions to the existence of alternatives. A new cosmopolitanism emerges in which it is impossible to maintain traditions 'in the traditional way'. So, along with globalization comes 'de-traditionalization.' Freed from the constraints of localism and traditionalism, both individuals and institutions become more 'reflexive': more self-conscious, and consequently more open to revising their practices and identities. Instead of a life whose main outlines are determined by the contingencies of birth - class, sex, locality - we are increasingly required to turn our lives into a 'reflexive' process of flexibly inventing and re-inventing our identities. Traditional forms of gender relation and family forms, established authority relations and norms of conduct and especially the traditional political divisions of left and right, rooted in traditional class identities, are expected to dissolve in the acid of reflexivity.

Beck and Giddens concur in their expectation that the mass politics of left and right, like the class identities which that expressed, will fade away, to be replaced by a new politics 'beyond left and right'. Giddens speaks of this as a politics of life-style and voluntary activity, whilst Beck's hope is for a 'new modernity' in which non-institutionalized social activism will demand democratic accountability from technocrats and politicians in the way science and technology are developed and introduced.

The ecological modernizers remain, however, significantly more optimistic about the environmental prospects of 'modernity' than either the analysts of 'risk society' or the radical political economists. In favour of the ecological modernizers is the evidence that the 'advanced' industrial societies have made significant progress in environmental regulation and 'green' taxation, most have ministries devoted to environmental policy, significant gains have been made in combating important sources of air, water and soil pollution, recycling, materials substitution, and increasingly energy and resource-efficient technologies have been developed and employed. Evidence on materials and energy flow over a twenty-year period for some of the industrialized countries does indeed show the looked-for 'decoupling' of economic growth measured in financial terms from measurable environmental impact: industry in these countries does appear to be increasingly ecologically efficient per unit of economic value produced.

In the domain of environmental politics, the green movements in most advanced industrial societies have changed their role from a marginal, oppositional and 'outsider' status, to insiders, collaborators with business, government and technocrats in setting mainstream policy objectives. In the international sphere, the EU has gained democratic legitimacy for its vigorous espousal of environmental issues, both in relation to the wider global scene and in relation to the record of member states. At the global level, a series of conferences leading up to Rio in 1992 have provided an overarching concept of sustainable development embracing both social justice and long-term environmental protection, as well as international agreements on, among other important issues, trade in endangered species, climate change, ozone depletion and conservation of biological diversity.

There is a major debate about how effective these agreements are, and indeed about what effectiveness means. International agreements do not operate in isolation from each other and they frequently have negative side-effects on other environmental problems. For instance, some substitutes for CFCs are powerful greenhouse gases, so the Montreal Convention on ozone-depleting substances has side-effects on the Kyoto Protocol, eventually leading to international action. Because of such interconnections and side-effects the real issue is whether the *system* of international environmental agreements promotes *sustainability* on balance. There is evidence that it does do so, pushing countries beyond what they would otherwise have done to promote sustainability.

As the constructionists clearly demonstrate, the formation and transformation of environmental issues as an agenda for public attention and policy-making depends on complex interactions between social movement activists, researchers, media communicators, policy networks and communities, industrial lobbies, government departments, international organisations and many other sorts of actors. In the realist approaches favoured here, recognition of the roles played by these heterogeneous and often conflicting social actors has to be complemented by acknowledgement of the active causal role played by non-human beings, relations and forces: both those purposively mobilized in the course of technologically mediated human social interaction with nature, and those unintentionally and often unexpectedly 'striking back'. Scholars have an important place in the effort to understand the systemic connections between the social, economic, political and biophysical dimensions of our increasingly problematic 'metabolism' with non-human nature. The intellectual demands of such an enterprise, and the great divisions of interest and of value judgement at stake in it suggest that it will always be a thoroughly contested enterprise. The less encouraging aspect of our situation, however, is that the socio-ecological processes of destruction and degradation escalate as we argue.

References

Barrett, S. and Graddy K. (2000) Freedom, growth and the environment. *Environment and Development Economics* 5(4):433–456.

Bell, M. M. (2004) *An Invitation to Environmental Sociology,* 2nd edn. Thousand Oaks: Pine Forge Press.

Brown, L. (2004) *Outgrowing the Earth.* New York: W. W. Norton

Midlarsky, M. (1998) Democracy and the environment: an empirical assessment. *Journal of Peace Research* 35(3): 341–361.

Mol, A.P.J. (2001) *Globalization and Environmental Reform: The Ecological Modernization of the Global Economy.* Boston: MIT Press.

Schnaiberg, A. (1980) *The Environment, From Surplus Scarcity*. New York: Oxford University Press.

Uphoff, N. (ed.) (2002) *Agroecological Innovations*. London: Earthscan

2

Environmental Sustainability

Most people in the world today have an immediate and intuitive sense of the urgent need to build a sustainable future. They may not be able to provide a precise definition of "sustainable development" or "sustainability" - indeed, even experts debate that issue - but they clearly sense the danger and the need for informed action. They smell the problem in the air; they taste it in their water; they see it in more congested living spaces and blemished landscapes; they read about it in the newspapers and hear about it on radio and television. The stories that carry the message may be about pollution alerts or the bans on driving and closed beaches that result from them, or about hunger and famine, growing health problems such as asthma and allergies, unsafe drinking water, "greenhouses gases" and the threat of global warming and rising ocean levels, the destruction of the world's forests and the expansion of its deserts, the disappearance of species, the large-scale death offish and birds caused by oil spills and pollution, or about forest fires, floods, dust storms, draughts and other so-called "natural" disasters. Or they may be about many other matters suggesting increasing levels of distress and desperation: inexplicable violence and outbreaks of war, mass migrations, the rise of intolerance and racism, the denial of democratic freedoms, corrupt practices that enrich the few at the expense of the many, rising prices and resource scarcities, growing unemployment and slipping standards of living for many of the world's inhabitants.

Even banal accounts of everyday life cause one to pause and reflect: e.g. the need for poor women in developing countries to spend many hours each day searching for wood and water or the plight of the commuter in the developed world whose travel to work is now measured in hours rather than minutes. Are these random events and developments, unrelated to one another, or are they parts of a pattern? Both opinion polls and casual conversations suggest that people are increasingly beginning to sense that something has gone seriously wrong, that there must be some connection among the growing difficulties they encounter and read about, even if they cannot provide an adequate explanation of exactly what it is or how it has developed.

What is "Sustainability"?

For countless millions of people, it is not simply a matter of speculating about causalities; they are already feeling the painful consequences of changing conditions in their daily lives. The situation is most severe for the poor and deprived, but increasingly even the more fortunate are experiencing a growing sense of anxiety and unease, of pending problems and unresolved difficulties. They perceive, for example, that the opportunities open to their children are shrinking even as the problems and challenges facing their societies continue to grow. Providing employment for the young and social security for the old is a growing challenge as populations expand in the developing world and age in the industrialized countries. Yet, if the future looks increasingly problematic, there is no retreating into the past, into time-honoured practices and values, for these, too, have been irreversibly altered by the profound transformation of society in the wake of the industrial and post-industrial revolutions. Attempts to do so usually end in frustration and sometimes in violence and nihilism.

What lies behind these newspaper headlines and the rising concerns and problems of people around the world? There is evidently no single or easy answer, but it is not difficult to identify some important and interrelate contributing factors:

- the rapid growth of the world's population and its changing distribution,
- the persistence of widespread poverty,
- the growing pressures placed on the environment by the worldwide spread of industry and the use of new and more intensive forms of agriculture,

- the continuing denial of democracy, violations of human rights and the rise of ethnic and religious conflicts and violence, gender inequity, and
- the very notion of "development" itself, what it has come to mean and how it is measured.

These factors are probably more accurately regarded as symptoms rather than as causes as they themselves are the consequences of thinking, values and practices in social, economic and political affairs that have put the world on an "unsustainable" path. It is, therefore, not only necessary to deal with the problems discussed below, but even more essential to get our thinking right: to see the interrelations among these problems and recognize the fundamental need to develop a new perspective rooted in the values of sustainability. It is this need which makes education the key to creating a sustainable future.

Population

In 1950, the estimated population of the planet was 2.5 billion. By the year 2000, it is projected to grow to more than six billion and by the year 2025 to more than eight billion. Population pressures are undeniably a factor in the degradation of environments. Moreover, population pressures are increasing most rapidly in the regions of the world where resources for coping with the requirements and demands of growing numbers are most limited. Between 1990 and 1995, an estimated 94% of total population growth occurred in the less developed regions and only 6% in the more developed regions. While fertility rates are declining in all major regions of the world in response to improved access to education, health and social services, especially by women, rapid population growth is projected to continue well into the 21st century. By even the most optimistic scenarios, the world's population will nearly double before it stabilizes. The predictable consequences of such growing population" pressures, especially in already densely populated and poor countries, include rapid urbanization, possible further reductions in living standards, lower *per capita* investments in education and health and increased environmental distress and degradation. Less predictable outcomes might include a rise in violence or even war, large-scale migrations and escalating poverty and famine. While many of the worst consequences could possibly be avoided by early preventive action, the record of past decades provides little support for optimism or complacency in this regard.

Poverty

Poverty is, in part, a consequence of the present pattern of population growth as well as a serious threat to both human dignity and sustainable development. Over a billion people, about a third of the total population of the developing countries, are desperately poor, struggling to survive on less than a dollar a day. Hundreds of millions more live on the threshold of poverty and face the constant risk of sinking below it. People unable to care for themselves can be excused for failing to care for their environment. Necessity drives them to use, and eventually overuse, all the resources at hand: land, water, wood, vegetation and, indeed, anything that can help them to meet their vital needs. Poverty also makes it difficult to mobilize people to work together for common goals, be they healthful environments, food security, jobs or other aspects of sustainable development. Poverty makes the delivery of education and health services more difficult and spurs population growth. Poverty also contributes to much of violence and war which destroy lives and undermine social and economic progress.

The solution to poverty must be found not only through economic measures, but also through political and social reforms, as poverty is caused not only by natural scarcities, but also by domination, exploitation and exclusion. Nor can the answer to poverty be sought only through increased production. What is produced, and the employment generated in producing it, must also be more equitably shared. At present, one quarter of the world's population consumes three-quarters of the world's natural resources. For particular resources, petroleum for example, the disparities are even greater: the average consumption of a North American is fifteen times greater than that of an Indian. At the extremes of wealth and deprivation - where the comparison is between individuals, not countries - the disparities defy belief: by one estimate, for example, the wealth of the world's richest 359 individuals equals the annual income of the poorest 2.4 billion people, almost 40% of humankind. Poverty reduction is, at once, an essential goal and indispensable condition for sustainable development.

Environmental Degradation

The rise of a worldwide industrial civilization during the past century and the parallel development of more intensive forms of agriculture - based on the use of chemical fertilizers, insecticides and herbicides - needed to feed and clothe a rapidly expanding population has placed unprecedented pressures and stresses on the world's ecosystems. The problems and dangers are manifold. In industrialized regions, the combustion of fossil fuels, the "life blood" of modern civilization, is resulting in an acidification of soils which is having a destructive impact on plants, forests and aquatic life in lakes and rivers. The use of fossil fuels is also responsible for the buildup of "greenhouse gases" that are a key factor in global warming which is changing weather patterns and raising ocean levels around the world. Even a modest rise in the earth's average temperature, of say two to three degrees, would result in the inundation of vast amounts of fertile low-lying coastal lands and the disappearance of many islands. A growing dependence on chemicals, many of which have never existed in nature, is having a damaging impact on plants and animals, leading even to the extinction of certain species and thus to a reduction in the world's biodiversity. Chemicals are affecting the stratosphere, depleting ozone and exposing the earth's surface to higher levels of ultra-violet radiation known to cause skin and other cancers. In the developing countries, land degradation presents perhaps the most immediate and urgent problem. As agricultural land *per capita* diminishes as a result of population growth and urbanization, it becomes essential to preserve the productivity of every available acre.

Yet, perhaps the greatest environmental danger lies in problems that are little discussed, even among scientists, such as the impact of human activities on natural "nutrient cycles" required to produce and balance elements essential to life: including carbon, oxygen and nitrogen. While the long-term affects of anthropogenic activities on the environment are either unknown or poorly understood, it is abundantly clear that delicate balances are being disturbed and disrupted. Some of these changes - e.g. the extinction of species - are already irreversible. Other processes may soon pass the point of no return, if action is not taken promptly. Yet, measures to protect the environment are resisted by those who insist that the needs of development - rising living standards for growing numbers - must take precedence over ecological concerns. The challenge of sustainability involves reconciling and adjudicating conflicting claims and moving towards a development which is environmentally sound.

Democracy, Human Rights and Peace

It is not only the harm that human beings are doing to nature, but also the injury that they are inflicting upon one another that is at the root of many of the world's problems. While democracy has made impressive progress in many regions, it has faltered in others. Even many "democratic countries", however, do not systematically practice democracy. The votes of the citizens may be counted, but the citizens themselves don't count for much in the operation of society. Inequalities dominate in all sphere of life: in the sharing of wealth, jobs, opportunities and social services, gender discrimination, and, of course, political influence and power. Underdevelopment and poverty are, at once, a cause and a consequence of arbitrary and undemocratic rule. The State, whose duty it is to protect the rule of law, is often the first to disregard it. Human rights are violated and ethnic and religious tensions exacerbated for political or personal gain. The bitter experience of the last decades has been that the failure of development efforts has often been followed by outbreaks of violence and wars between cultural, ethnic and religious communities. Since 1990, more than nine out of ten wars have broken out within countries rather than between them and more than nine out ten casualties have been civilians rather than soldiers. Without peace, there can be no development in any meaningful sense of the term.

Development

"Development" itself - what it means and how it is measured - is also an important part of the problem. Standard measures of development, such as gross national product, equate "development" with growth in production and consumption of goods and services. While such measures take into account investment in the means of production, such as the digging of copper mines or the drilling of oil wells, they fail to account for the use and eventual exhaustion of the precious capital represented by the world's endowment of natural resources. Nor, until quite recently, have economists adequately recognized that the capabilities embodied in women and men through education, experience and training are, in fact, the most essential "means of production".

Many economic measures fail on other scores as well. While they carefully account for productive outcomes, they treat the emission of smoke, gases and other pollutants, not as costs, but simply as "externalities". This is so because society as a whole, not the polluter, bears the cost and burden of coping with the problem. National accounts also fail to reflect what is done out of love or duty rather than for profit, thereby deeply discounting the indispensable work that women have always done - and continue to do

- for their families and in and about their homes. In addition, the focus of economics on the "immediate present" is seriously at odds with the need to consider the long-term well-being of the environment. In addition, while economic costs are viewed as incremental and linear, the impact of economic activity on the environment is cumulative and subject to sudden and possibly irreversible changes. The traditional visions of the economist and the ecologist are thus fundamentally at odds. Fortunately, a search for common understanding is underway.

Yet, perhaps the greatest problem arises from the automatic equation of higher levels of production - and by implication, consumption - with "development". Economists, and everyone else as well, recognize that this is at best a half truth. What is produced and, especially what it is used for, is every bit as important as how much of it is turned out. An added dollar of consumption, which doubles the daily income of an impoverished individual, evidently serves a very different purpose than the negligible addition of a dollar of purchasing power to the income of a millionaire. Yet, the automatic equation of a single technical measure of "development", usually GNP, with the society's overall progress and well-being is pervasive. It is part of an overall 20th century mind set that means are more important than ends, levels of activity more important than the purposes served.

Increasingly, developing nations set themselves the goal of "catching up" with Europe, Japan or the United States in levels of GNP *per capita.* It is necessary, however, to reflect that for all countries to achieve the current level of production of the most industrialized countries, worldwide consumption of natural resources would have to increase three-fold. Comparable increases would occur in the emission of pollutants and other perverse effects of production, assuming the newly industrializing countries make the same investment in controlling emissions as is presently done in the most industrialized countries. It may, however, be unrealistic to expect them to do so while urgent social needs are still unmet. The challenge is to find means and measures that assist the developing countries to meet the basic needs of their people without inflicting irreversible damage on their environment.

In pursuing this objective, measures of development such as the UNDP's Human Development Index, which seeks to take account of the many dimensions of human well-being, would be extremely helpful in focusing attention on the ends that development must serve rather than on means, such as increased production, alone.

Interdependence

None of the factors discussed above can be examined or acted upon in isolation from the others. They are in constant interaction. Violence, for example, is, at once, a cause of poverty and its consequence. Growing populations place increasing stress on ecosystems, but human activity by contributing to climate change further intensifies population pressures though desertification and rising ocean levels. The issues, moreover, are not only related to one another in a physical manner, but also in a psychological sense. How people think about the issues - their knowledge, beliefs, attitudes and especially their values - is as important in the search for solutions as the "objective realities" confronted.

It is also essential to take account of the varying circumstances in which people around the world find themselves and the impact of their situation upon their priorities and values. To an individual living in rural poverty in the developing world, "sustainable development", if it is to make any sense, must mean increased consumption and a higher living standard. By contrast, to an individual in a wealthy country, with a closet full of clothes, a pantry full of food and a garage full of cars, "sustainable development" could mean more modest and carefully considered consumption. Similarly, the issue of inter-generational parity and justice, inherent in discussions of sustainable development, will raise very different questions and choices in a country with a rapidly growing population, nearly half of which is under twenty years of age, than in a country with a stable and aging population.

In sum, the puzzle of sustainable development cannot be solved by concentrating on the pieces. It has to be seen as a whole - in both its scientific and social dimensions - not as a series of isolated issues and problems. In the final analysis, sustainable development is humanity's response to an emerging global challenge and crisis.

North-south Differences

Any discussion of sustainable development has to take account of both the disparities between rich and poor nations and, equally important, what these differences imply for policy formulation. Policies appropriate to the north make little sense for the south and vice versa. The major challenge facing the developing nations of the south is that of significantly increasing productivity and output to overcome poverty and deprivation. Until this is achieved, at least in part, it is unrealistic to expect the environmental standards that apply in the north to be adopted in the south as well. So long as poverty is widespread, the south will be tempted - and often compelled - to accept higher levels of pollution and lower levels of control, just as the highly industrialized countries of today did until relatively recently.

One of the tasks of education for sustainable development, especially in northern countries, will be to explain why these differences exist and why, at least for the time being, the application of equal standards would result in very inequitable outcomes. The north can, of course, help to shorten the interval in which lower standards will be necessary by assisting the south in its development efforts, especially, as concerns education, in the development of its schools, universities and training programmes for scientists and other key personnel through financial and technical assistance, sharing of knowledge, and the training of experts from the south in their institutes and universities. There are, of course, many similarities as well as differences. Many of the emerging issues - e.g., the need for renewal energies and fresh water - affect all regions of the world in varying degrees. A major challenge of education for sustainable development in all countries is that of helping people to understand and adapt to a pace of change which is, as yet, "unnatural" to all cultures. In a deeper sense, as neighbours on the same planet our destinies are ultimately joined.

Towards Sustainability

Sustainable development has been variously defined and described. It is not a fixed notion, but rather a process of change in the relationships between social, economic and natural systems and processes. The World Commission on Environment and Development, for example, defined sustainable development in terms of the present and the future: "Sustainable development is development that meets the needs of the present without compromising the ability of future generations to meet their own needs". Other definitions have extended the notion of equity between the present and the future, to equity between countries and continents, races and classes, genders and ages.

Perhaps the most widely used definitions focus on the relationship between social development and economic opportunity, on the one hand,

and the requirements of the environment on the other: i.e., on improving the quality of life for all, especially of the poor and deprived, within the carrying capacity of supporting ecosystems. This does not necessarily set fixed limits on "development", but rather recognizes that the prevailing notions and definitions of development must themselves evolve in relation to changing requirements and possibilities. *Caring forthe Earth* (IUCN, 1991), for example, notes that a sustainable economy "can continue to develop by adapting, and through improvements in knowledge, organisation, technical efficiency and wisdom". In brief, sustainability calls for a dynamic balance among many factors, including the social, cultural and economic requirements of humankind and the imperative need to safeguard the natural environment of which humanity is a part. What is sought is the condition of "human security" for all people.

A Dynamic Balance

Sustainability, in effect, involves an equation between environmental requirements and development needs. It can be balanced by acting either to reduce stresses or to increase "carrying capacities". The argument between ecologists and economists has been that the former have stressed the first course of action and the economists the latter. It is evident that in a crisis, both possibilities have to be carefully explored. There are environmental strains, such as the multiplying of populations, that at some point become incompatible with both the maintenance of the environment and the quality of life. By the year 2030, for example, it is projected that there will be three billion more people on earth than today. The task of feeding, clothing and sheltering them will be enormous, that of providing them with education, employment, security and a minimum of well-being and satisfaction vastly greater still. These facts of life must not be ignored. But neither should the capacity of humanity to find and invent solutions be overlooked or minimized.

The higher levels of production required by three billion additional people will certainly inflict serious damage upon the environment unless modes of production change significantly in the coming decades. Fortunately, this is what is happening. Within ten to twenty years, the introduction of automobiles, buses and trucks powered by such fuel cells is expected to substantially improve air quality in large cities of industrialized countries. Whether the cost of this new technology will be affordable in the developing regions of the world in the near future is of central importance and, as yet, unclear. Of greater relevance to developing countries, major breakthroughs are being made in agriculture which allow farmers to produce more food on less land while reducing the impact on the environment. Moreover, the widening use of computers and the growing reach of the new information and communication technologies are ensuring a far more rapid and wider dissemination and application of innovations than was the case even a decade ago.

While such developments are highly encouraging, it would be imprudent to expect science and technology to find a solution to every problem that humanity is capable of creating for itself. Nor would it be wise to rely on technical solutions alone without considering the capacity of human societies to adjust to the changes and stresses that they may impose. But it would be equally short-sighted to overlook the capacity of people to invent solutions to problems or to find ingenious ways of coping with such problems. The concept of sustainable development is informed by both the warnings of environmentalists and the arguments of economists in favour of development. It seeks to strike a realistic balance between dangers and possibilities, hopes and fears, aspirations and constraints. The "point of balance" is, of course, influenced by many factors and, thus, subject to constant change.

An Emerging Vision

Yet, while there are many definitions of sustainable development, it can perhaps be better understood as an emerging vision rather than as a neatly defined concept or relationship. In truth, it is as much an ethical precept as a scientific concept, as concerned with notions of equity as with theories of global warming. Sustainable development is widely understood to involve the natural sciences and economics, but it is even more fundamentally concerned with culture: with the values people hold and how they perceive their relations with others. It responds to an imperative need to imagine a new basis for relationships among peoples and with the habitat that sustains human life.

Its strength is that it frankly acknowledges the interdependence of human needs and environmental requirements. In so doing, it rejects the single-minded pursuit of one objective at the cost of others. A heedless pursuit of "development", for example, can not be accepted at the cost of inflicting irreparable damage on the environment. But neither can the preservation of the environment be achieved at the cost of maintaining half of humanity in poverty. Or, in the terms in which the debate is sometimes posed, we cannot sacrifice people to save elephants, but neither can we - at least not for very long - save the people by sacrificing the elephants. Indeed, this is a false dichotomy that must be rejected. We must imagine a new and sustainable relationship between humanity and its habitat: one that places humanity at centre stage, but does not neglect that what is happening in the "wings" may turn the drama of everyday life into an ancient Greek tragedy in which we see a terrible fate approaching, but can muster up neither the collective will nor common means to escape it.

Education for the Future

It is widely agreed that education is the most effective means that society possesses for confronting the challenges of the future. Indeed, education will shape the world of tomorrow. Progress increasingly depends upon the products of educated minds: upon research, invention, innovation and adaptation. Of course, educated minds and instincts are needed not only in laboratories and research institutes, but in every walk of life. Indeed, access to education is the *sine qua non* for effective participation in the life of the modern world at all levels. Education, to be certain, is not the whole answer to every problem. But education, in its broadest sense, must be a vital part of all efforts to imagine and create new relations among people and to foster greater respect for the needs of the environment.

Education must not be equated with schooling or formal education alone. It includes non-formal and informal modes of instruction and learning as well, including traditional learning acquired in the home and community. By defining education broadly, one also widens the community of educators, as the programme statement of *Education 21* promoted within the United Kingdom notes, to include "teachers, lecturers, curriculum developers, administrators, support staff, industrial trainers, countryside rangers and staff, environmental health and planning officers, education officers with NGOs, community educators, youth leaders, parent association members, media people, representatives of learners in all contexts - and yet more." One might further widen this community to include all those, whatever their role in society, who perceive a need or duty to inform and educate people regarding the requirements of a sustainable future. International organizations, government departments and institutions, foundations and many others are deeply involved in education in the broad sense of the term used here. Many firms in the private sector also seed the need to play their part in promoting awareness and are doing so in innovative ways: for example, through sponsoring the publication of articles in newspapers and journals exploring environmental and social issues. This vast community of educators represents an enormously potent, but largely untapped human-resource for sustainable development that can be invaluable in a range of contexts as well as education. It represents, above all, a means for bringing the struggle for sustainable development into communities and local institutions around the world which, in the final analysis, is where the cause of sustainable development will either triumph or fail..

Education serves society in a variety of ways. The goal of education is to make people wiser, more knowledgeable, better informed, ethical, responsible, critical and capable of continuing to learn. Were all people to possess such abilities and qualities, the world's problems would not be automatically solved, but the means and the will to address them would be at hand. Education also serves society by providing a critical reflection on the world, especially its failings and injustices, and by promoting greater consciousness and awareness, exploring new visions and concepts, and inventing new techniques and tools. Education is also the means for disseminating knowledge and developing skills, for bringing about desired changes in behaviours, values and lifestyles, and for promoting public support for the continuing and fundamental changes that will be required if humanity is to alter its course, leaving the familiar path that is leading towards growing difficulties and possible catastrophe, and starting the uphill climb towards sustainability. Education, in short, is humanity's best hope and most effective means in the quest to achieve sustainable development.

Public Awareness and Understanding

Awareness is a prelude to informed action. In democratic societies, action towards sustainable development will ultimately depend on public awareness, understanding and support. Common information and shared understandings, however, are important not only for mobilizing public support, but also for carrying out work consultative and participatory approaches in all fields.

Public awareness and understanding are, at once, consequences of education and influences on the educational process. A public well informed of the need for sustainable development will insist that public educational institutions include in their curricula the scientific and other subject matters needed to enable people to participate effectively in the numerous activities directed towards achieving sustainable development. The students that emerge from such courses will, for their part, be alert to the need for public authorities to make adequate provision for the protection of the environment in all development plans. Education is particularly important in developing a "taste for knowledge".

Perhaps the greatest problem that advocates of sustainable development face is to convince not only those who are opposed to their ideas, but also those who simply "don't want to know". An approach that emphasizes local issues, rather than global ones, is likely to be most effective in dealing with this constituency. This may account, in part, for the success of non-formal community education and local environmental communication programmes in reaching and sensitizing people to environmental and development issues in both developing and industrialized countries. A particular benefit of such programmes is that they are often directly linked to action to control or solve the problems identified.

Effective Communication

Advocates of sustainable development and the environmentalists who proceeded them have learned much about how to communicate effectively. Initially, their emphasis was on "getting the science right" with little thought given to how to communicate findings and make them meaningful to a wide and non-technical public. It was assumed that the facts would speak for themselves. Sadly, it didn't prove that easy. It is important to explore the difficulties that arose in order that they may be avoided in the future. There are several sorts of problems: the influence of vested interests, the neglect or inadequacy of communication strategies, the complexity of the messages and the unfortunate tendency of some of the messengers to spend more time squabbling with one another than communicating with the public.

Confronting Vested Interests

In any struggle - including one to win over the minds of the public - it is important to understand the motives and strengths of those on the other side of the issue. Naively, one might imagine that few would find reason to oppose measures necessary to avoid potentially calamitous consequences for humanity. But, alas, what is good for humanity in general may nonetheless be costly and inconvenient to particular individuals, groups and other vested interests. The electrical power industry, to cite a current example, is vigorously against more stringent controls on the emission of "greenhouse gases", even though there is convincing and growing evidence - if not yet certainty - that the build-up of such gases in the atmosphere is leading to global warming and all that could ensue from it. Regulation is not going to come about on the basis of the evidence alone. Public mobilization and vigilance are essential, if effective measures are to be enacted into law and enforced. Until quite recently, advocates of the common interest have had difficulty mustering the needed public relations expertise and support to overcome the influence of vested interests. Fortunately, in the past two decades, many lessons have been learned, especially by environmentalists, on how to convert a growing public concern for the state of earth into effective support for specific measures to address concrete problems. Yet, in most countries, while environmental issues are now receiving greater support, measures aimed at promoting population policies, social development, poverty reduction and other necessary measures for achieving sustainable patterns of development continue to be largely ignored by the general public. Ultimately, though, there can be no solution to environmental problems unless the social and economic ills besetting humankind are seriously addressed. It is this broader message and reality which remains to be effectively communicated to and internalized by the public.

Debate and defence of particular interests are, of course, inherent in the democratic process. Vested interests have to be overcome by democratic means: namely, by more effective mobilization of public opinion aimed at gaining support at all levels: international, national and local. The difficulties in achieving this goal should not, however, be under-estimated. As can be seen in the discussions about implementation of the Convention on Climate Change, the opposition comes not only from particular industrial interests, but also from countries and groups of countries. While nobody favours pollution per se, many countries would nonetheless like to exempt themselves or others from bearing the cost of stringent controls. A vigilant and informed world public represents a powerful counterweight to the vested interests that appear, at present, to have the upper hand on many issues. It is no accident that the countries that are militating most strongly for controls on emissions and other environmental measures are the same nations that have strong environmental lobbies and publics committed to action - locally, nationally and internationally - to preserve the environment.

Communication Strategies

One of the lessons of recent experience is the need to establish effective communication strategies as an integral part of any major scientific inquiry or programme. A comparison, which highlights this need, can be made between the National Acid Precipitation Assessment Project (NAPAP) in the United States and the Intergovernmental Panel on Climate Change (IPCC) established in 1988 by the United Nations Environmental Programme (UNEP) and the World Meteorological Organization (WMO). NAPAP, while it was highly regarded by scientists, had virtually no communication strategy. As a result, while its research and recommendations were well considered, there has been little follow-up action. IPCC has sought to avoid this failure by keeping both the scientific community and the general public systematically informed of its work and findings from the very start. It has also sought, with considerable success, to build bridges to policy-makers in order that they be aware of and able to reflect upon the implications of the panel's emerging findings. By informing the public, IPCC has made it much more difficult to simply sweep its conclusions under the carpet. The lesson here is that communication has to be seen as a long-term interactive process strategically aimed at particular groups and audiences, not as a concluding message when a project or panel is about to present its final report and wind up its activities. It is not necessary - or even desirable - for scientists to become propagandists, but it is essential that studies conducted in the public interest have adequate means to communicate their findings to the public on whose behalf they were carried out.

Complex Messages

The messages of sustainable development represent a challenge in and of themselves. Rather than being simple and unambiguous - thus easy to communicate - environmental and developmental issues tend to be complex. This is so because of the inherent complexity of ecological and human systems. They defy simplistic explanations, solutions and predictions. Some scientists, for example, expect that the build-up of greenhouse gases that causes global warming may, initially, result in a several decades of falling temperatures in particular regions of the world because of the affect of the melting of the polar ice caps in slowing or stopping the warm ocean currents This contention may or may not be correct. There is great uncertainty about what may happen, even if there is a general consensus that it won't be favourable to life on earth. Yet, the ambiguity of the situation makes it hard to explain to non-specialists. To the general public, hot and cold are opposites, even if to the climate scientists they are merely different manifestations of environmental stress. To urge people to beware of global warming, but to keep both their woollens and their beachwear handy, just in case, is in no way convincing. Such uncertainty suggests that global warming may be more speculative than scientific. This, evidently, is not the case. The truth is that complex realities are difficult to communicate in simple messages. Yet, attempts to simplify what, by its very nature is not simple, may result in further confusion and misunderstandings and, ultimately, in lack of credibility.

The same problems arise, although to a lesser degree, in dealing with major transformations such as population growth and urbanization. For example, the projection that by the middle of the next century several cities in the developing world may have populations approaching, or even exceeding, 50 million may be accepted matter-of-factly without due reflection on what is involved in managing an urban centre on such a scale or what the quality of life might be for its inhabitants. Thus, while the statement may appear easily understandable, the problems and issues that it raises may go undetected or be seriously under-estimated. The fact is that people have difficulty adjusting from the scale of things encountered in everyday life to the scales of magnitude - enormously large and infinitesimally small - needed to understand demographic or ecological phenomena. Ultimately, a solution can be found only by educating the public in the developmental and environmental "facts of life". Indeed, in the 21st century, the literacies of science, ecology and development will be as essential to comprehending the world as were the traditional skills of reading and writing at the start of the present century.

In the meantime, it will be important for those advocating sustainable development to choose, wherever possible, those cases and examples that are most easily understood by the general public. For example, air pollution is, if not always visible, often capable of being smelled and tasted. Closed beaches are plausible evidence of the pollution of rivers, lakes and seas. Even if one wishes, or is compelled, to go on to discuss complex issues such as global warming, it is well to begin with the evidence at hand: car, bus and truck exhaust, smoke stacks, etc. Health issues also are readily understood by the general public: allergies, asthma, and bronchial infections are widely accepted as consequences of deteriorating air quality. Average citizens may not gain a perfect understanding of global warming from such evidence - even scientists don't have that -but their common sense will tell them that what harms their environments and health is capable of doing even greater damage on a global scale. The basic dictum of pedagogy is to begin where the learner is. This is also good advice for the communication specialist. Start with problems that people feel and understand at the local level. That is both valuable knowledge in itself and, if need be, a basis for moving on to more complex and global understandings.

The Messengers

Emotionalism and exaggeration are another frequent source of difficulty. The press is understandably drawn to those with the most extreme views: e.g., to ecological fundamentalists who will accept no compromise or individuals who possess an apocalyptic vision of the future. Disagreements and quarrels between specialists are also "newsworthy" and are skillfully exploited by the opponents of reform proposals to suggest that the evidence supporting them is weak and confused. More moderate and reasoned voices often go unheard in the din. Extreme positions, while they may be useful in catching the public's attention and alerting it to pending dangers, make it difficult to move from declarations and debate into action.

It has to be recognized that neither individuals nor societies are ready or even able to change their habits and behaviours from one day to the next. Proposals for change, if they are to be effective, have to be feasible. Both the messages and the messengers have to appear credible and responsible. Nothing is to be gained by scaring people. Alarmist predictions that make it seem that the world is about to end are evidently not conducive to the long-term planning and action that sustainable development requires. On the contrary, it is far more effective to present problems as manageable through responsible conduct and, wherever possible, put forward a realistic solution and a means to take preventive action.

Reason for Optimism

Yet, while these problems are important and need to be addressed, there is abundant reason for optimism. People are becoming increasingly concerned about the crises afflicting the environment and impeding development. This is, in part, because communication on such issues has become more effective and focused, but mainly because these problems - especially those concerned with the environment - are increasingly impinging on people's lives: their health, their comfort and their hopes for the future. Scientific data alone have rarely won an argument when people were not ready to accept its conclusions and, equally rarely, have scientific findings lost an argument in which people had an intuitive sense that the data were right and relevant. The climate of opinion is changing and becoming more favourable to the promotion of sustainable development. This is not an opportunity to be wasted, but a chance to be seized. To do so will require effective leadership - not in the sense of direction from above - but in the form of responsibility and responsiveness. There is also an important role for the press and media in responding and building upon the growing interest of the public in sustainable development concerns. Evidently, the fullest possible use must be made of the new information and communication media, but traditional and folk media must also be creatively employed. The messages of ancient cultures on sustainable development often took the form of metaphors and analogies. These are still powerful means of communicating, especially with poor peoples who are often little schooled, but very much in tune with their culture. The greatest challenge is precisely that of reaching the more than one billion people who live in poverty and deprivation, often in remote rural regions, urban slums and refugee camps.

Reorienting Education to Sustainability

Until recently, the planet was a large world in which human activities and their effects were neatly compartmentalized within nations..and within broad areas of concern (environmental, economic, social). These compartments have begun to dissolve. This applies in particular to the various global 'crises' that have seized public concern.. These are not separate crises: an environmental crisis, a development crisis, an energy crisis. They are all one."

Reorienting education to sustainability requires recognizing that traditional compartments and categories can no longer remain in isolation from each other and that we must work increasingly at the interface of disciplines in order to address the complex problems of today's world. This is true both within education, where interdisciplinarity is slowly and with difficulty gaining ground, and between the spheres of education, work and leisure as lifelong learning emerges as a key concept for planning and developing educational systems. It is also true as concerns the most important boundary of all: that separating those included in education systems from those who are excluded from them. These changes are not occurring nearly as rapidly as would be desired, but they are nonetheless taking shape within education at all levels.

Ultimately, sustainable development will require an education that not only continues throughout life, but is also as broad as life itself, an education that serves all people, draws upon all domains of knowledge and seeks to integrate learning into all of life's major activities. The time when education was the activity of childhood and work the pursuit of adults is long over. The rapid growth of knowledge has rendered the notion of schooling as a "once and for all" preparation for life utterly obsolete. The growth of knowledge is advancing exponentially, yet not nearly as fast as the need for understanding and solutions at which it is aimed. As concerns sustainable development specifically, it is impossible to predict with reliability what will be the key issues on which people will need information in five, ten, twenty or fifty years. It is predictable, however, that such developments will not fit neatly into the existing and artificial sub-divisions of knowledge which have been in place for more than a century. Hence, understanding and solving complex problems is likely to require intensified co-operation among scientific fields as well as between the pure sciences and the social sciences. Reorienting education to sustainable development will, in short, require important, even dramatic changes, in nearly all areas.

The importance of education was underscored at the at the nineteenth Special Session of the General Assembly convened to review the implementation of Agenda 21 five years after Rio. The resolution adopted by the session emphasized that a "fundamental prerequisite for sustainable development is an adequately financed and effective educational system at all levels, particularly the primary and secondary levels, that is accessible to all and that augments both human capacity and well- being... Even in countries with strong educational systems," the resolution continues, "there is a need to reorient education, awareness and training to increase widespread public understanding, critical analysis and support for sustainable development. Education for a sustainable future should engage a wide spectrum of institutions and sectors..and should include the preparation of sustainable development education plans and programmes".

Importance of Basic Education

Inherent in the concept of sustainability is the vision of a more equitable world. This can only be achieved by providing the disadvantaged with the means to advance themselves and their families. And of these means, the most essential is education, particularly basic education. Over 100 million children between the ages of 6 and 11 never attend school and tens of millions more enter school only to drop out within a few months or years. Moreover, there are over 800 million illiterate adults, most of whom have never been enrolled in school. The first requirement in the quest for development and equity must be to change this situation and make schooling of quality available to all. But that goal, alas, is still far off. For the present, the challenge is to make the best of an unfortunate and unjust situation.

The World Conference on Education for All used the term "basic education" to refer to all forms of organized education and training that meet the basic learning needs of individuals, including literacy and numeracy, as well as the general knowledge, skills, values and attitudes that they require to survive, develop their capacities, live and work in dignity, improve the quality of their lives, make informed decisions and continue learning. The Conference consciously choose to define education in terms of learning outcomes rather than levels of instruction.

Given the situation today in many developing countries, it does not suffice to orient formal education towards sustainability. Attention also has to be given to those who are presently unserved or poorly served by schools. This is a large group, well over a billion people, and a vital one for the future. Girls and women, the mothers of today and tomorrow, are in the majority. They are, or will be, the first and most influential teachers of their children. The goals of educating young children are focused on ensuring their health, development, happiness, well-being and adjustment to the environment in which they live. If these goals are not achieved, the future of the child is compromised and the prospects of sustainable development diminished.

Basic education provides the foundation for all future education and learning. Its goal, as concerns those in the pre-school primary school-age population, whether enrolled in school or not, is to produce children who are happy with themselves and with others, who find learning exciting and develop inquiring minds, who begin to build up a storehouse of knowledge about the world and, more importantly, an approach to seeking knowledge that they can use and develop throughout their lives. Basic education is aimed at all the essential goals of education: learning to know, to do, to be (i.e., to assume one's duties and responsibilities) and to live together with others, as outlined in *Education: the Treasure Within,* the report of the Independent Commission on Education for the Twenty-first Century Report published in 1996 by UNESCO. It is, thus, not only the foundation for lifelong learning, but also the foundation for sustainable development.

Basic education for adults is aimed at empowerment. It is, in the words of the Amman Affirmation, the document summing up the mid-decade review of progress towards Education for All, "..the key to establishing and reinforcing democracy, to development which is both sustainable and humane and to peace funded upon mutual respect and social justice. Indeed in a world in which creativity and knowledge play an ever greater role, the right to education is nothing less than the right to participate in the life of the modern world." In sum, if our vision of the future is a world based on democracy, striving to achieve greater social justice and economic opportunity, and concerned to improve the quality of life and preserve the environment, then basic education has to be the first order of business, for it holds the power to contribute to all of these goals by enabling people to take their destinies into their own hands and play their role in shaping the common destiny of humanity. Sustainable development cannot be achieved by a small minority on behalf of the vast majority. It will require the contribution and commitment of each and all. That is why it is essential to give all people the means - starting with basic education - to participate in shaping a sustainable future.

Curriculum Reform

To advance such goals, a curriculum reoriented towards sustainability would place the notion of citizenship among its primary objectives. This would require a revision of many existing curricula and the development of objectives and content themes, and teaching, learning and assessment processes that emphasize moral virtues, ethical motivation and ability to work with others to help build a sustainable future. Viewing education for sustainability as a contribution to a politically literate society is central to the reformulation of education and calls for a "new generation" of theorizing and practice in education and a rethinking of many familiar approaches, including within environmental education.

Education for sustainability calls for a balanced approach which avoids undue emphasis on changes in individual lifestyles. It has to be recognized that many of the world's problems, including environmental problems, are related to our ways of living, and that solutions imply transforming the social conditions of human life as well as changes in individual lifestyles. This draws attention to the economic and political structures which cause poverty and other forms of social injustice and foster unsustainable practices. It also draws attention to the need for students to learn the many processes for solving these problems through a broad and comprehensive education related not only to mastery of different subject matters, but equally to discovering real world problems of their society and the requirements for changing them.

This kind of orientation would require, *inter alia,* increased attention to the humanities and social sciences in the curriculum. The natural sciences provide important abstract knowledge of the world but, of themselves, do not contribute to the values and attitudes that must be the foundation of sustainable development. Even increased study of ecology is not sufficient to reorient education towards sustainability. Even though ecology has been described by some as the foundation discipline of environmental education, studies of the biophysical and geophysical work are a necessary - but not sufficient -prerequisite to understanding sustainability. The traditional primacy of nature study, and the often apolitical contexts in which is taught, need to be balanced with the study of social sciences and humanities. Learning about the interactions of ecological processes would then be associated with market forces, cultural values, equitable decision-making, government action and the environmental impacts of human activities in a holistic interdependent manner.

A reaffirmation of the contribution of education to society means that the central goals of education must include helping students learn how to identify elements of unsustainable development that concern them and how to address them. Students need to learn how to reflect critically on their place in the world and to consider what sustainability means to them and their communities. They need to practice envisioning alternative ways of development and living, evaluating alternative visions, learning how to negotiate and justify choices between visions, and making plans for achieving desired ones, and participating in community life to bring such visions into effect. These are the skills and abilities which underlie good citizenship, and make education for sustainability part of a process of building an informed, concerned and active populace. In this way, education for sustainability contributes to education for democracy and peace.

Structural Reform

Reorienting the curriculum towards sustainable development requires at least two major structural reforms in education. The first is to re-examine the centralized mandating of courses and textbooks in order to allow for locally relevant learning programmes. Local decision-making can be facilitated through the reform of centralized educational policies and curricula, and the formulation of appropriate syllabuses and assessment policies. Nationally-endorsed syllabuses can serve as "broad framework documents" which provide aims and general objectives for subjects, an overview of broad content themes, appropriate learning experiences, relevant resource materials, and criteria for assessing student learning. This type of syllabus can provide centralized accountability, while allowing schools, teachers and students to make choices about the specific learning experience, the relative depth and breadth *of* treatment *for* different topics, the case studies and educational resources used, and how to assess student achievements.

A second major area of structural reform is the development of new ways to assess the processes and outcomes of learning. Such reform should be inspired by what people want from their educational system, as well as what society needs. The period of profound change in which we are living needs to be taken into account by educational systems, which were, for the most part, designed to serve a society which is fast becoming history. Learning needs to be seen as a lifelong process which empowers people to live useful and productive lives. The reorientation of education along these lines - and in anticipation to the extent possible of future needs - is fundamental for sustainable development, including its ultimate objective not only of human survival but especially of human well-being and happiness. Similarly, there also needs to be a revamping of the methods of credentialing students. The various ways in which students are judged (testing, report cards, evaluations) and the basis for awarding diplomas at all levels need to reflect the reformulation of outcomes of learning towards sustainability.

Educational Reform: A Case Study

What does reorienting education towards sustainability mean in practical terms? This is the question that educators immediately want to know. Does it mean adding courses to an already overweight curriculum? Will it require new teaching approaches and methods? New physical facilities, equipment and textbooks to be purchased from an already severely pinched budget? Is it something that can be achieved in a month, a school year or several years? While the reform of education in this direction is still more talked about than put into practice, there are examples emerging which shed light on how to move in this direction.

One example is that of the Toronto (Canada) Board of Education which recently undertook a reform of its curriculum through a massive community consultation. Thousands of parents, students, staff and members of the public contributed to full day community consultations aimed at exploring how education should respond to the demands of a changing world. The focus of the inquiry was the question "What should students know, do and value by the time they graduate from school?". Although the notion of "sustainability" was not imposed, it emerged as an essential requirement in the course of the consultation.

The education that parents and the community wanted for their children was in many respects hardly revolutionary or even surprising. The six graduation outcomes specified were: literacy; aesthetic appreciation and creativity; communication and collaboration; information management; responsible citizenship; and personal life skills, values and actions. These differ from most traditional curricular objectives in that they are broader and more closely related to the needs and organization of life than to the requirements and structures of schooling.

The essence of the Toronto reform is that the curriculum is no longer focused exclusively on the traditional core subjects of language, mathematics, history, etc. Informed by the new vision of what the community felt tomorrow's students would need to know and be able to do, these disciplines underwent major revision. Mathematics, for example, now includes the skill of comprehending extremely large and extremely small numbers - e.g., ppm and ppb - which are essential to being environmentally literate and capable of understanding relative risk factors both in personal life and at work. Health now includes environmental issues including cancer, allergies and food additives as well as "consumerism".

Much of the success of the Toronto reform is due to the fact that it was not - and was not seen to be - an effort to change education to meet goals set by an elite or unduly influenced by outside pressures. The impetus to change came from within. The new curriculum had equal or greater academic rigour, but far greater relevance to life outside school walls. What it demonstrates is that education for sustainable development is simply good education, and that good education needs to make children aware of the growing interdependence of life on earth - interdependence among peoples and among natural systems - in order to prepare them for the future.

Toronto had one great advantage in implementing its curriculum reform: well-educated and well-trained teachers. In reality, what students learn is not necessarily what is written in the syllabus; it is what the teacher delivers in the classrooms. By far, the most frequent cause of curriculum failure is inadequate teacher training. In Toronto the development of the curriculum itself constituted an informal type of training in which thousands of teachers were involved. This was followed up by more formalized sessions and by systematic provision for teachers to upgrade their qualification through university courses and other forms of training. The lesson in this is that efforts to adapt education systems to sustainable development have to consider not only the question, "what are the essential messages that must be delivered?", but also and equally, that of "how will teachers be trained to put those messages across powerfully and effectively?".

In general, reforms aimed at sustainability will require much more of teachers than do traditional curricula. Students will have to be more actively involved in individual and collective activities. This will require teachers to play new roles which, in turn, implies a need for increased training and support. Educational reforms, like the movement towards sustainable development itself, requires holistic and systematic thinking; piecemeal approaches will not suffice and can not produce the required results.

Reform at Different Scales

It must, of course, be recognized that curriculum reform can take place in different ways and on different scales. If schools are granted greater autonomy, as proposed above, significant reforms could take place within schools or even classrooms, rather than at the national, provincial or district levels. Certain of these reforms would be aimed at changes in particular lessons or courses rather than for the curriculum as a whole. Such reforms would not be sufficient to fully orient the curriculum towards sustainability, but they could nonetheless be highly valuable. It is also necessary to recognize that schools and school systems in many developing countries are struggling under enormous burdens. They have insufficient resources to implement their present programmes of study - often only four or five textbooks are available for a class of fifty or more - and no means to aim at the more ambitious objectives that were possible in Toronto and in other industrialized countries. This inequality in educational resources and, hence in opportunities, is itself one of the major causes of unsustainability. If schools are to be a means to the reform of society, it is then essential that society at all levels - local, national and international - invest adequate attention and resources in its schools.

Contribution of Environmental Education

It is clear that the roots of education for sustainable development are firmly planted in environmental education. While environmental education is not the only discipline with a strong role to play in the reorienting process, it is an important ally. In its brief twenty-five year history, environmental education has steadily striven towards goals and outcomes similar and comparable to those inherent in the concept of sustainability.

In the early 1970s, the emerging environmental education movement was given a powerful boost by the United Nations Conference on the Human Environment held in Stockholm in 1972, which recommended that environmental education be recognized and promoted in all countries. This recommendation led to the launching in 1975 by UNESCO and the UN Environment Programme (UNEP) of the International Environmental Education Programme (IEEP), which continued until 1995. The influence of the IEEP - and the national and international activities which it inspired - has been widely felt and is reflected in many of the educational innovations carried out in the last two decades.

That work was inspired largely by the guiding principles of environmental education laid down by the Intergovernmental Conference on Environmental Education held in Tbilisi in 1977, which followed a comprehensive preparatory process which included the International Workshop on Environmental Education held in Belgrade in 1975 to draft the concepts and vision which were later taken up by governments in Tbilisi. These encompass a broad spectrum of environmental, social, ethical, economic and cultural dimensions. Indeed, the recommendations of the Rio Conference, held fifteen years later, echo those of Tbilisi, as is evident in the following quotations from the report of the 1977 conference:

"A basic aim of environmental education is to succeed in making individuals and communities understand the complex nature of the natural and the built environments resulting from the interaction of their biological, physical, social, economic and cultural aspects, and acquire the knowledge, values, attitudes and practical skills to participate in a responsible and effective way in anticipating and solving environmental problems, and the management of the quality of the environment."

"A further basic aim of environmental education is clearly to show the economic, political and ecological interdependence of the modern world, in which decisions and actions by the different countries can have international repercussions. Environment should, in this regard, help to develop a sense of responsibility and solidarity among countries and regions.."

"Special attention should be paid to understanding the complex relations between socio-economic development and the improvement of the environment."

These principles were successfully translated into educational goals and, with greater difficulty, into schoolroom practice in many countries.

The motto of the environmental education movement has been: "think globally, act locally". Over a period of more than two decades, it developed a highly active pedagogy based on this premise. In the early grades, in particular, the emphasis was upon learning the local environment through field studies and classroom experiments. By starting in the primary grades, before the process of compartmentalization that marks secondary and particularly higher education sets in, students were encouraged to examine environmental issues from different angles and perspectives.

The influence of environmental education in promoting interdisciplinary inquiries can be seen at all levels of education. A course on environmental economics, for example, looks to anthropology for insights and source material. It studies the decline of ancient civilizations - e.g., the Sumerian and Mayan and that of the Easter Islands - which exploited their environment without due regard for its sustainability. Equally valuable lessons might be drawn from tribes and groups that faced challenging environmental conditions, but survived against difficult odds by developing an awe, love and respect for nature. In many such cultures, the environment was placed in the sphere of the sacred and used according to a set of well-defined rules that, whatever their origins, served to prevent the over-use and exhaustion of natural resources. Environmental education has also found original ways of looking at and measuring human impact on the environment, such as the "ecological footprint", which estimates the number of acres of land required to sustain individuals according to their lifestyles and patterns of consumption. Innovative work has also been done in the field of environmental health by relating illness to environmental stress and ways of life.

In brief, the record of the environmental education movement is one of resourcefulness, innovation and continuing accomplishments. Lessons learned from environmental education provide valuable insight for developing the broader notion of education for sustainable development.

Interdisciplinarity

A basic premise of education for sustainability is that just as there is a wholeness and interdependence to life in all its forms, so must there be a unity and wholeness to efforts to understand it and ensure its continuation. This calls for both interdisciplinary inquiry and action. It does not, of course, imply an end to work within traditional disciplines. A disciplinary focus is often helpful, even necessary, in allowing the depth of inquiry needed for major breakthroughs and discoveries. But increasingly, important discoveries are being made not within disciplines, but on the borders between them. This is particularly true in fields such as environmental studies which are not easily confined to a single discipline. Despite this realization and a broadening support for interdisciplinary inquiries, the frontiers between academic disciplines remain stoutly defended by professional bodies, career structures and criteria for promotion and advancement. It is no accident that environmental education and, more recently, education for sustainable development, has progressed more rapidly at the secondary and primary levels than within the realm of higher education.

Yet, higher education has an indispensable role to play. This is true both in the area of research and in the training of specialists and leaders in all fields. A failure to develop educational programmes related to sustainability in universities and specialized institutes has, therefore, an impact on society as a whole. It is, for example, increasingly important to include appropriate materials on sustainable development in the programmes of study of journalists, engineers, managers, doctors, lawyers, scientists, economists, administrators and numerous other professions. Universities could also render a valuable service by building components of sustainable development into the special programmes for teachers, senior managers, local leaders such as mayors, parliamentarians and others in leadership positions. Universities also play a key role in international cooperation and would do so more effectively it they gave fuller consideration to the needs of scientists and social scientists from developing countries, especially as concerns interdisciplinary inquires into environment and development issues.

Fortunately, the situation appears to be evolving in a favorable direction. Faculties of economics, for example, are adding specialists in environmental economics to their staffs who, by the very nature of their work, are required to develop expertise in scientific and other disciplines. As students arrive at the university from secondary schools with experience in and a taste for interdisciplinary work, universities in many countries are slowly adapting to meet their needs and demands. Major research projects, such as that on climate change, are also developing in specialists the habit of working across disciplines. Ultimately, the growing necessity for interdisciplinary inquires can be expected to reduce the resistance imposed by entrenched habits and conservative institutional structures.

Shifting to Sustainable Lifestyles

The effectiveness of awareness raising and education for sustainable development must ultimately be measured by the degree to which they change the attitudes and behaviors of people, both in their individual roles, including those of producers and consumers, and in carrying out their collective responsibilities and duties as citizens. Both of these roles - the private and the public - are indispensable and mutually reinforcing. A willingness to change one's lifestyle or pattern of consumption is essential but insufficient in itself to bring about societal change. Similarly, one's action as a citizen, no matter how responsible and far-sighted, is inadequate if one's individual behavior and lifestyle are wasteful or destructive. Sustainable development requires both individual enlightenment and responsibility and appropriate policies and action by public authorities and the private sector. If, for example, an individual wishes to use public transportation to reduce urban congestion and pollution, this choice can be effective only if such transport exists and, for most people, will become habitual only if the it is both economic and convenient. In short, the move toward sustainable lifestyles is not merely a matter of individual choice; it also requires collective action and responsibility.

Individual Choices and Lifestyles

The caretaking consumer insists upon purchasing products that are kind to the environment. While this choice is often regarded as a virtuous one and as a burden on the consumer, that need not be the case. It is, for example, possible to develop automobiles that would provide the same comforts as present models, but would go twice as far on a litre of fuel. Production close to home - the site of consumption - is also more energy efficient and, in the case of food, offers greater freshness as well. The resistance to "green choices" often comes not from consumers, but from producers who must change their modes of operation and make substantial investments in new means of production and distribution. Yet, ultimately, the producer must provide what the consumer demands. In this way, alert and informed consumers can exercise an influence on what is produced and sold.

Eco-efficiency, i.e. increasing the added value of our activities while reducing our use of resources and the impact on the environment, may have an important role to play in safeguarding the environment. Eco-efficiency calls for better management of existing processes or products to reduce waste, use less energy and facilitate reuse and recycling. New and alternative technologies may also allow for cleaner production. Brazil, for example, is one of several countries that has established training programmes for industrial designers aimed at creating more efficient products. There is also a growing tendency for ecologically responsible firms to assume responsibility for a product throughout-life-cycle: selling the product, maintaining it and recycling it when it is no longer needed. A more radical approach is to look at the underlying need that a consumer meets through consumption, rather than at the means which are currently used to meet that need. For example, the primary use of a car is to provide mobility. Providing suitable, reliable alternatives - such as public transport, car-sharing schemes, home shopping or interactive communications - may enable consumers to meet their needs while having a lower impact on the environment. Eco-efficiency may thus entail a switch in thinking from the supply of products to the supply of services.

All of these notions are included in the concept of "sustainable consumption" and "sustainable lifestyles". Sustainable consumption does not necessarily mean consuming less. It means changing unsustainable patterns of consumption by allowing consumers to enjoy a high quality of life by consuming differently. For example, consumers might be encouraged to buy products which have a longer life, which can be easily repaired or updated. Policies to influence consumption patterns need to work with the market and recognize consumers' needs and desires, thereby helping consumers make informed choices. Here, too, education has an important part to play.

Obviously, sustainable consumption has a different meaning when applied to the poor in developing countries and to the socially-excluded in developed countries. For the poor, sustainable development is not only about preserving the environment, but also, even mainly, about providing for human needs such as food, clothing, shelter, health care and security. However, even in the poorest countries, sustainable consumption means avoiding waste. The special session of the United Nations General Assembly on the follow-up to the Earth Summit emphasized the need for developing countries to promote sustainable consumption patterns in the development process.

To sum up, the individual through wise consumer choices and responsible behavior can certainly play a useful role in reducing waste and favoring products that do less harm to the environment. These are valuable and practical contributions, but also symbolic ones. The willingness of citizens to sort their trash for recycling, for example, sends a message to elected officials that voters are concerned about the environment.

Collective Decision-making

The engaged citizens of a democratic society can exercise a strong influence on behalf of sustainable development through their civic role as well as through their behavior as consumers and producers. Individual lifestyles are inevitably strongly conditioned by public policies. Sorting trash, for example, serves little purpose if the collection service then throws everything into a dump. Individual action must be complemented and supported by public policies at all levels.

The main instruments of public policy are laws and regulations, tax and fiscal policies (i.e. the use the government makes of public revenues), and social instruments, especially education. Regulation is the most direct approach: the production and use of certain toxic chemicals or other environmentally harmful products are being controlled and phased out in many countries as less dangerous substitutes are developed. Tax policies, including subsidies, are highly effective in influencing relative prices. Ecological tax reform usually consists in heavily taxing environmentally harmful products and favoring products that are kinder to the environment through reductions in taxes or through subsidies, designed to make such products more competitive. Norway, for example, is moving towards a revenue-neutral tax reform in which taxes on harmful products are used to subsidize "green" products and services. An important part of such reforms is to correct the negative incentives in present tax codes. In the United States, for example, it is estimated that the total subsidy to automobiles - through building and maintenance of roads and other measures - amounts to some $300 billion dollars annually. In European countries, by contrast, the automobile is a major source of state revenues. This difference is reflected in fuel prices: a litre of fuel in many European countries costs more than a gallon (nearly four times as much) in the United States.

Finally, governments and public entities such as schools and hospitals have enormous purchasing power. If procurement policies favor the purchase of environmentally friendly products, producers come under enormous pressure, both directly in the marketplace and indirectly through banks, insurance companies and others who are fearful of the consequences and liabilities they might incur.

In sum, in democratic societies public policy responds to the will of the people. It is here that public awareness and understanding of the need for sustainable development best expresses itself through support for laws, regulations and policies favourable to the environment. People express their preferences as they decide how to spend their money, as well as through the ballot. Public action, through voting or otherwise, is contingent on more than "public awareness". What is needed is an understanding of the issues, and of the likely consequences of a given purchasing or electoral decision. For this reason, growing attention has been given in Europe and elsewhere to "ecological audits" and labelling which allows producers of environmentally benign products to advertise this fact to the public. Electoral choices are, of course, effective in terms of sustainability only to the extent that the public is well informed. Public understanding is the foundation for people to fulfill their roles as responsible citizens, consumers and public-spirited individuals.

Sustainability as a Moral Imperative

Sustainable development calls for comprehensive change in the way society operates. Production and consumption must be restructured in ways that better meet the basic needs of all in an ecologically responsible manner. The present disparities between rich and poor - unimaginable wealth at one extreme and desperate poverty at the other -have to be reduced. Population growth must be moderated and ecologically unsound practices quickly reduced and eventually eliminated. These steps call not only for practical action, but also for fundamental changes in perceptions and values, indeed, for a renewal of culture that will enable societies to confront the major challenge of the 21st century: the quest for sustainable development.

Ethical values are the principal factor in social cohesion and, at the same time, the most effective agent of change and transformation. Achieving sustainability will depend ultimately on changes in behavior and lifestyles, changes which will need to be motivated by a shift in values and rooted in the cultural and moral precepts upon which behavior is predicated. Without change of this kind, even the most enlightened legislation, the cleanest technology, the most sophisticated research will not succeed in steering society towards the long-term goal of sustainability. Education in the broadest sense will by necessity play a pivotal role in bringing about the deep change required, in both tangible and intangible ways.

Some Ethical Principles of Sustainability

Like the notion of sustainability itself, the ethics of sustainability can not be defined or detailed simply. In fact, thinking along these lines in just emerging, as we struggle to combine existing notions to develop a new, integrated ethical concept for the future. For the moment, it is possible to indicate a few new lines of thinking in terms of principles which associate values such as human rights and responsibility, intergenerational equity, solidarity, justice, democracy, freedom of expression, and tolerance.

The "Ethic of Time"

The last few years have been marked by a thorough assessment of the worrisome trends of our times and the negotiation of international, regional and national action plans to redress these trends before they become irreversible. We have at hand, therefore, the diagnosis and in many cases the cure to the ills of the world. What is needed is to take corrective action before it is too late, action which may be costly or unpopular, but which must transpire nevertheless. The need to improve the world is obviously not new. What is new is the risk of irreversibility and all that it implies for the future of society as we know it. The notion of an "ethic of time" is the moral imperative to take action before reaching the point of no return.

Within this notion of time, there is also that of "anticipation" and "prevention". Those in a position to effect change - governments, international organizations, scientific institutions and universities, business and industry - need to shake off inertia and the temptation to find short-term half-way solutions. They need to take up the challenge of doing what all agree needs to be done. While this may seem self-evident, the assessment of progress made five years after the Rio Conference has revealed that in fact little real change has taken place. Despite all efforts and good intentions, we have so far been unsuccessful in coming to terms with the future. The balance sheet of Rio + 5 is harsh proof of the futility of shallow responses to complex and grave problems. The question is: How long can we wait to adopt a new ethic for the future, an ethic which will drive us to rectify our current path and to anticipate our future needs, regardless of how broad or how deep the required changes need to be.

Complexity as an Ethical Issue

Problems related to sustainable development are characterized, *inter alia,* by their complexity. This complexity must be communicated and understood, even though to do so is not easy or necessarily palatable. The simplification of complex issues - so often observed today - is not only fraudulent in that it misrepresents reality, but also irresponsible on the part of those who understand these issues. It is here that the scientific and intellectual communities bear a particular moral responsibility, to ensure that decision-makers as well as the public are fully cognizant of the multiple dimensions of the problems they face. To the extent that knowledge will be the driving force of change, the link between ethics and science will be key to solving many of the problems of the future.

Continuity: The Ethical Link between Past, Present, and Future

Each action "today" is a step towards creating "tomorrow": this has always been true. However, never before has the weight of today - with the full cognizance of those "in charge" - been so determinant of humanity's future. Nothing less than the viability of our planet is at stake. Humanity is thus in a position of power as well as responsibility not only towards the people alive on Earth today, but also towards generations yet unborn who will have no choice but to accept the reality which we will have created. In considering the ethics of sustainability, our moral responsibility towards future generations is of primordial importance. In living up to this responsibility, we must strive to achieve balance and continuity between meeting the needs of today without compromising those of the future, and without losing the memory of what history has already taught us. Recognizing the intergenerational dimension of sustainability is, of course, not new: the report of the Brundtland Commission marked the beginning of thinking of development in terms of the future as well as the present. What is still lacking is a meaningful adjustment of our way of life to reflect this ethical imperative.

In November 1997, UNESCO adopted a Declaration on the Responsibilities of Present Generations Towards Future Generations, which aims to ensure through action today a viable future for the coming generations. The Declaration is the fruit of a cooperative effort undertaken since 1994 by Commandant Cousteau's team and UNESCO, involving researchers and government experts before the final adoption by UNESCO Member States. Its preamble recalls certain fundamental principles: "the necessity for establishing new, equitable and global links of partnership and intra-generational solidarity [...] the avowal that the fate of future generations depends to a great extent on decisions and actions taken today and that present-day problems, including poverty, technological and material underdevelopment, unemployment and exclusion, discrimination and threats to the environment, must be solved in the interests of both present and future generations".

The 12 articles of the Declaration put forward proposals on what can be done to safeguard the needs and interests of future generations in the fields of education, science, culture and communication. Concerning the environment, for example, Article 4 states that "the present generations have the responsibility to bequeath to future generations an Earth which will not one day be irreversibly damaged by human activity. Each generation inheriting the Earth temporarily shall take care to use natural resources reasonably and ensure that life is not prejudiced by harmful modifications of the ecosystems and that scientific and technological progress in all fields does not harm life on Earth." The idea is reinforced in Article 5 which stipulates that the present generations "should ensure that future generations are not exposed to pollution which may endanger their health or their existence itself." Emphasizing the importance of culture, the Declaration considers it the responsibility of the present generation to "identify, protect and safeguard the tangible and intangible cultural heritage and to transmit this common heritage to future generations" (Article 7). This is also the thrust of the articles concerning development and biodiversity. There is, on the one hand, the questions of ensuring "the conditions of equitable, sustainable and universal socioeconomic development" (Article 10) and, on the other, of protecting the "human genome, in full respect of the dignity of the human person" (Article 6).

Culture and Sustainability

While culture is elusive to definition, it may be taken to refer to all those mentally generated forms of organization created, preserved and transmitted within a social group or, in a wider context, the human species. Culture includes our whole system of beliefs, values, attitudes, customs, institutions and social relations. It shapes the way we perceive the world (including ourselves) and how we interact with it. To the extent that the global crisis facing humanity is a reflection of our collective values, behavior and lifestyles, it is, above all, a cultural crisis.

Culture is, therefore, an inextricable part of the complex notion of sustainability. It can be seen as an arbiter in the difficult trade-offs between conflicting ends with regard to development goals. As pointed out in the report of the World Commission on Culture and Development set up jointly by UNESCO and the United Nations, culture is not only the "servant of ends but (..) the social basis of the ends themselves", a factor of development but also the "fountain of our progress and creativity".

For example, changing wasteful patterns of consumption, particularly in the industrialized countries, is an area where culture will clearly have an instrumental role to play. Changes in lifestyle will need to be accompanied by a new ethical awareness whereby the inhabitants of rich countries discover within their cultures the source of a new and active solidarity which will make it possible to eradicate the widespread poverty which now besets 80% of the world's population as well as the environmental degradation and other problems which are linked to it.

For all the people of the world, culture is a very practical, concrete determinant of sustainable development. The kind of change required by sustainability implicates each community, each household, each individual. Successful solutions to problems at this level of society will need to be rooted in the cultural specificity of the town or region if the people are to be supportive of and involved in such change.

And yet the cultural diversity of humankind is today in jeopardy. In this respect, a parallel can be drawn between biological diversity and cultural diversity, which may be seen as aspects of the same phenomenon. Just as the multitude of diverse species and life forms that constitute the Earth's biological diversity have evolved to adapt to different geographical and climatic conditions, so the adaptability of the human species is expressed in humanity's cultural diversity. Just as nature produces a variety of species adapted to their environment, so humankind develops varied cultures in response to local conditions. Cultural diversity may thus be seen as a form of adaptive diversity and, as such, a prior condition to sustainability. The present trend towards globalization is threatening the richness of human cultures, and has already destroyed many traditional cultures. The argument for halting the loss of species is also applicable to the loss of cultures, and the consequent diminishing of humanity's collective repertoire for survival.

Towards a Common Ethic

Today people are more aware than ever of global realities. We are beginning to understand the impact of our individual and collective actions on ourselves and on the biosphere as a whole. The concept of sustainability is in itself a reflection of this new awareness. Perhaps we are beginning to move towards a new global ethic which transcends all other systems of allegiance and belief, which is rooted in a consciousness of the interrelatedness and sanctity of life. Would such a common ethic have the power to motivate us to modify our current dangerous course? There is obviously no ready answer to this question, except to say that without a moral and ethical foundation, sustainability is unlikely to become a reality.

Education, Ethics, and Change

In the early 18th century, the bases of the industrial revolution that was then beginning did not exist: neither mentally, nor socially nor technically. But the vision of a society organized in a new way and operating by new rules took root. At the close of the 20th century, it is clear that societies are beginning to consider the concept of sustainable development and, in some cases, to confront the profound changes that it implies. Fundamental social changes, such as those required to move towards sustainability, come about either because people sense an ethical imperative to change or because leaders have the political will to lead in that direction and sense that the people will follow them. Human societies are skillful at estimating risks, dangers and limitations. They are much less experienced in calculating their own potentialities: their capacities to invent, innovate, discover, reorganize, create, correct and improve. Societies need to be convinced of the need for sustainable development, in order to show their capacity to devise solutions to the problems confronting them.

It is in this context that education an public awareness are seen as essential to bringing about conditions conducive to sustainable development. Ethical values, such as equity, are shaped through education, in the broadest sense of the term. Education is also essential in enabling people to use their ethical values to make informed and ethical choices. Over time, education also powerfully affects cultures and societies, increasing their concern over unsustainable practices and their capacities to confront and master change. Indeed, the potential of education is enormous. Not only can it inform people, it can change them. It is not only a means for personal enlightenment, but also for cultural renewal. Education not only provides the scientific and technical skills required, it also provides the motivation, justification, and social support for pursuing and applying them. Education increases the capacities of people to transform their visions of society into operational realities. It is for this reason that education is the primary agent of transformation towards sustainable development. It is also for this reason that society must be deeply concerned that much of the education presently on offer falls far short of what is required. Improving the quality and coverage of education and reorienting its goals to recognize the importance of sustainable development must be among society's highest priorities.

Mobilizing for Action

While sustainability is a long-term goal for human society and a process which will necessarily need to take place over time, there is a sense of urgency to make progress quickly before "time runs out". We are therefore faced with a tremendous challenge, a challenge of unprecedented scope, scale and complexity. We are pressed to act even as we are still working out new concepts and new methodologies. We are pushed to change structures and mindsets, yet there is no obvious path, no model which shows the way. Experimentation and innovation are the watchwords, as we search - often simply through trial and error - for adequate solutions. And we must do all this in a climate of sweeping economic, social and political change, while being exhorted to "do more with less".

An International Framework for Action

We do have an internationally negotiated framework for action which has been hammered out during the series of United Nations conferences dealing with different aspects of sustainable development, beginning in 1992 with Rio (environment and development), and followed in 1994 by Cairo (population), in 1995 by Copenhagen (social development) and Beijing (women), and in 1996 by Istanbul (human settlements). Each of these conferences, as well as the three conventions on biological diversity, climate change, and desertification, contain explicit recommendations or whole chapters devoted to education and public awareness. The international consensus which these agreements represent is a solid and comprehensive basis for moving forward.

At the heart of this new international consensus is a new vision of education, public awareness and training as the essential underpinning for sustainable development, a linchpin to support advances in other spheres, such as science, technology, legislation, and production. Within the action plans, education is no longer seen as an end in itself, but as a means to:

- bring about the changes in values, behavior and lifestyle that are needed to achieve sustainable development, and ultimately democracy, human security and peace;
- disseminate knowledge, know-how and skills that are needed to bring about sustainable production and consumption patterns and to improve the management of natural resources, agriculture, energy and industrial production;
- ensure an informed populace that is prepared to support changes towards sustainability emerging from other sectors.

These action plans are to be implemented not only for international institutions such as the United Nations system, but also and most importantly by national and local entities. A range of "major groups" - including women, youth, farmers, parliamentarians, scientists, business and industry and others - are called upon to participate, as well as governments and non-governmental organizations at all levels.

Chapter 36 of Agenda 21

Chapter 36 of Agenda 21, enitled "Promoting education, public awareness and training", provides an umbrella for all action related to education for sustainable development, including that foreseen by the other UN conferences. It is sweeping and comprehensive in scope. In addressing three programme areas (A. Reorienting education towards sustainable development; B. Increasing public awareness; C. Promoting training), chapter 36 encompasses all streams of education, both formal and non-formal, including basic education and environmental education.

After the Rio Conference, governments decided to establish within the United Nations the intergovernmental Commission on Sustainable Development (CSD) which meets each year to monitor the implementation of Agenda 21. Chapter 36 was reviewed in 1996 at the fourth session of the CSD, which decided at that time to adopt a special work programme which outlines priorities for action within the vast scope of chapter 36. Work programmes are initiated by the Commission for themes considered of particular importance and for which it wants some concrete action to be taken on its behalf. In addition to education, the CSD has work programmes on changing consumption and production patterns, technology transfer, and indicators for sustainable development.

At the nineteenth special session of the United Nations General Assembly held in June 1997 to review implementation of Agenda 21 five years after Rio, the work of chapter 36 was given further impetus by the reaffirmation of governments of the importance of education in achieving sustainability.

Action at the National Level

It is at this level that overall strategies for sustainable development must be given clarity and impetus and the need to integrate education into them in creative and effective ways recognized and acted upon. This involves the national government - which has the leadership role - major NGOs and associations, citizens groups, including corporate citizens, educational and other specialized institutions. Also involved are the agencies and organizations of the United Nations system which are endeavoring to work closely together with national authorities in implementing the recommendations of the various international conferences that have pointed the way towards sustainable development. The media should also be engaged to explain the purposes and goals of programmes and in making government plans and action known to the public.

What should be done at the National Level?

Governments should play a leading role in explaining the vision and benefits of sustainable development to the public. It should do so through all means available, public and private. As already noted, the media have an important role to play, including both the most modem and the most traditional media. The national school curriculum, at all levels, must incorporate the messages of sustainable development. These messages must also be emphasized in pre-service and in-service programmes of teacher training. Specialized institutions of all sorts should be actively involved. The relationship between environmental factors and health problems, for example, should be explained by doctors, nurses and hospitals.

Government leaders have to translate the concept of sustainable development into clearly definable steps and goals. Such goals should be set for every sector. There should, for example, be a clear timetable for reorienting the education system with the necessary budget and resources to achieve the goal. In many countries, national strategies or plans for sustainable development are instrumental in mobilizing and focusing efforts around national priorities. It is essential that national leaders demonstrate that there is political will to give priority to sustainable development and that they see public awareness raising and education and training as essential means for achieving national objectives.

The civil society at all levels, especially the national, should express its support for vigorous action aimed at advancing towards sustainable development. Organizations with specialized interests and competencies should not only support government action, but also monitor and assess them within their areas of competence and keep the public informed of both progress and problems. Teachers associations, for example, should carefully assess the progress being achieved in reorienting the education system towards sustainability and should keep both their members and the public-at-large informed of the situation. Increasingly, the United Nations system recognizes that the national level is where actions in support of sustainable development are most effective. The United Nations system also recognizes that to be effective in promoting such action it must work as a team, not as a collection of individual organizations and agencies, in close cooperation with national partners.

Action at the Local Level

The axiom "think globally, act locally" is as true today as ever. The role of the local community is of particular importance because the movement towards sustainable development can not succeed on a "top down" basis alone. What is needed is not only passive consent, but also active involvement on the part of the people. Actors at this level include the community and its leaders, local government, NGOs and associations working in the community, the private sector, local departments and services of national government, and of course people of all ages including youth. In a growing number of communities, local Agenda 21 groups have been established and are taking the lead in mobilizing support for local initiatives. Local needs will evidently determine local priorities and actions. It is important, however, for the local community, in consultation with national leaders and national government, to understand its place in the "big picture" of national and global action for sustainable development. The meaning and vision of sustainable development should be disseminated, discussed and debated in order to promote understanding and win community support. These discussions should take place in all community settings and institutions, including in schools.

The need for sustainable development at the local level must be understood. Practices that are not sustainable should be identified. Possibilities for correcting them can then be discussed and explored. It is critically important that the entire community, especially women, be involved in this discussion. Women, especially in rural communities, usually play a key role in the economic as well as in the social and cultural aspects of life. The local community and the household are important entry points for messages on sustainable development, especially for adults and out-of-school children. The schools, too, at all levels, should be actively involved in both discussions about and action to achieve sustainable development.

References

Benton, T. (1998) Sustainable development and the accumulation of capital: reconciling the irreconcilable? In Dobson A. (ed.) *Fairness and Futurity*. Oxford: Oxford University Press.

Brunkhorst, D., Bridgewater, P. and Parker, P. (1997) The UNESCO biosphere reserve program comes of age: learning by doing; landscape models for sustainable conservation and resource use. In Hale P. and Lamb D. (eds) *Conservation Outside Nature Areas*. University of Queensland, pp. 176–182.

Carney, D. (1998) *Sustainable Rural Livelihoods.* London: Department For International Development

Ward, H. (2006) International linkages and environmental sustainability: the effectiveness of the regime and IGO networks. *Journal of Peace Research* 43:149–166

3

Global Climate Change

The Earth has an atmosphere, consisting mostly of nitrogen (78%), oxygen (21%) and a number of greenhouse gases, which affect the balance of light and ultraviolet energy coming from the Sun and heat energy leaving the Earth, the global energy balance. The greenhouse gases, including carbon dioxide, methane, nitrous oxide and water vapour, absorb energy at some wavelengths of the electromagnetic spectrum, but allow energy at other wavelengths to pass through unimpeded.

Shorter wavelength sunlight passes through the atmosphere relatively unimpeded, although the ozone layer does absorb a lot of higher wavelength ultraviolet energy. This heats up the Earth's surface, causing it to re-release energy back to the atmosphere. The lower wavelength heat (or infrared) energy given off by the Earth, however, is mostly trapped (absorbed) by the greenhouse gases, and is prevented from escaping directly back into space. This trapped heat warms the Earth's surface on average by as much as 33°C. The process is called the natural greenhouse effect, and without it, the Earth would be as cold as the moon, about -18°C.

The atmosphere also contains millions of microscopic particles called aerosols, which scatter sunlight. The scattering of sunlight by the atmosphere is important because like the natural greenhouse effect, this can affect the amount of energy stored in the atmosphere, and therefore the Earth's climate. Changes in the composition of the atmosphere can be a mechanism of climate change.

A change in the greenhouse gas content of the atmosphere will affect the amount of energy stored in the atmosphere. For example, if the amount of carbon dioxide in the atmosphere is increased, more heat will be trapped in the atmosphere. This enhanced greenhouse effect raises the Earth's surface temperature. Changes in the amount of greenhouse gases in the atmosphere can occur in numerous ways. Natural changes in the amount of carbon dioxide stored in the atmosphere occurred at the end of the last Ice Age. A 50% increase in carbon dioxide levels coincided with a 5°C rise in global average surface temperature. Today, mankind through the burning of fossil fuels for energy and transportation, and changing land use, has produced a substantial change in the atmospheric composition over most recent centuries, and it is feared that this continuing change will lead to global warming.

The general state of the Earth's climate is dependent upon the amount of energy stored by the climate system, and in particular the balance between the amount of energy the Earth receives from the Sun, in the form of light and ultraviolet radiation, and the amount of energy the Earth releases back to space, in the form of infrared heat energy. Causes of climate change involve any process that can alter this global energy balance. Scientists call this "climate forcing". Climate forcing "forces" the climate to change.

The are many climate forcing processes, but broadly speaking, they can be separated into internal and external types. External processes operate outside planet Earth, and includes changes in the global energy balance due to variations in the Earth's orbit around the Sun, and changes in the amount of energy received from the Sun. Internal processes operates from within the Earth's climate system, and include changes in the global energy balance due to changes in ocean circulation or changes in the composition of the atmosphere. Other climate forcing processes include the impacts of large volcanic eruptions and collisions with comets or meteorites.

Luckily, the Earth is not hit by large comets or meteorites very often, perhaps every 20 to 30 million years or so, and therefore their associated climate changes occur rarely throughout Earth History. However, other causes of climate change influence the Earth on much shorter time scales, with changes sometimes occurring within a single generation. Indeed, our present pollution of the atmosphere with greenhouse gases may be causing the global climate to change. This man-made climate change has become known as global warming.

The Earth's climate changes when the amount of energy stored by the climate system is varied. The most significant changes occur when the global energy balance between incoming energy from the Sun and outgoing heat from the Earth is upset. There are a number of natural mechanisms that can upset this balance, for example fluctuations in the Earth's orbit, variations in ocean circulation and changes in the composition of the Earth's atmosphere. In recent times, the latter has been evident as a consequence not of natural processes but of man-made pollution, through emissions of greenhouse gases. By altering the global energy balance, such mechanisms "force" the climate to change. Consequently, scientists call them "climate forcing" mechanisms.

Climate System

The key to understanding global climate change is to first understand what global climate is, and how it operates. At the planetary scale, the global climate is regulated by how much energy the Earth receives from the Sun. However, the global climate is also affected by other flows of energy which take place within the climate system itself. This global climate system is made up of the atmosphere, the oceans, the ice sheets (cryosphere), living organisms (biosphere) and the soils, sediments and rocks (geosphere), which all affect, to a greater or less extent, the movement of heat around the Earth's surface.

The atmosphere plays a crucial role in the regulation of Earth's climate. The atmosphere is a mixture of different gases and aerosols (suspended liquid and solid particles) collectively known as air. Air consists mostly of nitrogen (78%) and oxygen (21%). However, despite their relative scarcity, the so-called greenhouse gases, including carbon dioxide and methane, have a dramatic effect on the amount of energy that is stored within the atmosphere, and consequently the Earth's climate. These greenhouse gases trap heat within the lower atmosphere that is trying to escape to space, and in doing so, make the surface of the Earth hotter. This heat trapping is called the natural greenhouse effect, and keeps the Earth 33°C warmer than it would otherwise be. In the last 200 years, man-made emissions of greenhouse gases have enhanced the natural greenhouse effect, which may be causing global warming.

The atmosphere however, does not operate as an isolated system. Flows of energy take place between the atmosphere and the other parts of the climate system, most significantly the world's oceans. For example, ocean currents move heat from warm equatorial latitudes to colder polar latitudes. Heat is also transferred via moisture. Water evaporating from the surface of the oceans stores heat which is subsequently released when the vapour condenses to form clouds and rain. The significance of the oceans is that they store a much greater quantity of heat than the atmosphere. The top 200 metres of the world's oceans store 30 times as much heat as the atmosphere. Therefore, flows of energy between the oceans and the atmosphere can have dramatic effects on the global climate.

The world's ice sheets, glaciers and sea ice, collectively known as the cryosphere, have a significant impact on the Earth's climate. The cryosphere includes Antarctica, the Arctic Ocean, Greenland, Northern Canada, Northern Siberia and most of the high mountain ranges throughout the world, where sub-zero temperatures persist throughout the year. Snow and ice, being white, reflect a lot of sunlight, instead of absorbing it. Without the cryosphere, more energy would be absorbed at the Earth's surface rather than reflected, and consequently the temperature of the atmosphere would be much higher.

All land plants make food from the photosynthesis of carbon dioxide and water in the presence of sunlight. Through this utilisation of carbon dioxide in the atmosphere, plants have the ability to regulate the global climate. In the oceans, microscopic plankton utilise carbon dioxide dissolved in seawater for photosynthesis and the manufacture of their tiny carbonate shells. The oceans replace the utilised carbon dioxide by "sucking" down the gas from the atmosphere. When the plankton die, their carbonate shells sink to the seafloor, effectively locking away the carbon dioxide from the atmosphere. Such a "biological pump" reduces by at least four-fold the atmospheric concentration of carbon dioxide, significantly weakening the Earth's natural greenhouse effect, and reducing the Earth's surface temperature.

Comets and Meteorites

The Earth's atmosphere protects us from the impacts of comets and meteorites, by vaporising all or most of the incoming material before reaching the Earth's surface. Scientists believe however, that every once in a while a celestial body of significant size collides with the Earth, causing untold destruction and an ensuing global climate change. The comet/

meteorite impact theory of climate change has been considered to account for the extinction of the dinosaurs 65 million years ago. Although critics of this theory suggest volcanism offers a better explanation for the mass extinction, evidence does exist which seems to support the case for a meteoritic impact which occurred at this time in Earth History.

A proposed impact site that has been dated at 65 millions years can be found on the tip of the Yucatan Peninsula in the Gulf of Mexico, Central America. In addition, a thin layer of clay deposited at this time at many localities around the Earth contains the rare element iridium. Although released in volcanic lava originating from deep within the Earth's interior, the impact theory indicates that the required iridium could come from a celestial object 10km across. An impact from a body this size would have released huge quantities of vaporised material into the atmosphere, blocking out the Sun and causing an initial "impact winter". Debris containing the iridium would gradually return to Earth forming the clay deposit. After this there would have been an increase in temperatures caused by the large amounts of carbon dioxide released by a spread of global fires. In addition, chemical reactions taking place in the atmosphere between pollutants would result in the formation of globally distributed acid rains.

Luckily such impacts only occur rarely, perhaps every few million or tens of millions of years. Although smaller objects hit the Earth more frequently, they have much less impact. Nevertheless, it is almost certain that another large comet or meteorite will at some time in the future strike the planet with potential consequences for the global climate and for life on Earth.

Continental Drift

The process by which the continents drift about the world is called plate tectonics. The movement of the Earth's plates, on which the continents ride, is very slow, being only a few centimetres each year. However, over tens or hundreds of millions of years, both the size and position of land areas can change appreciably. At times in Earth history, there have been super-continents in which all the continental plates were locked together in one area of the globe. The last of these occurred about 250 million years ago, and is named Pangea. Since that time, the continents have gradually moved apart, the most recent separation occurring between Europe and North America, during the last 60 to 70 million years, to form what is now the

North Atlantic Ocean. What is now the Pacific Ocean used once to be the vast expanse of water called the Panthalassa Ocean that surrounded Pangea.

Changes in the distribution of landmasses are believed to explain climate changes that occur over tens or hundreds of millions of years. Of course, we have no direct way of knowing what the Earth's climate was like hundreds of millions of years ago, but we can use geological records of sea floor sediments to reconstruct what the climate may have been like. We can also use computer models to estimate how different arrangements of continents may influence the global climate. Currently, it is believed that the arrangement of continental landmasses significantly affects the ocean currents. Since ocean circulation is involved in the transfer of heat around the Earth, so the wandering of landmasses over tens and hundreds of millions of years may influence climate changes over similar time scales.

Such long-term changes to ocean circulation as a result of continental drift may explain the gradual return to an ice-covered world during the last 40 million years. Prior to that period, there was little ice covering the polar regions. As the supercontinent of Pangea continued to break up, so the continent of Antarctica became isolated at the Earth's southern pole. With ocean all around it, a new circumpolar ocean current formed, and heat from lower latitudes was prevented from reaching the continent. The subsequent expansion of the (white) ice sheet on Antarctica about 35 million years ago increased the amount of sunlight the Earth as a whole reflected, and led to a drop in average global surface temperature. Today, although we live in a period of relative warmth since the end of the last Ice Age 14,000 years ago, the Earth as a whole is still gripped by a much longer period of global frigidity. Temperatures today are still perhaps 10°C cooler on average than they were during the age of the dinosaurs.

As continents break apart, new oceans form between them, through a process known as sea-floor spreading. A major zone of sea-floor spreading is today located along the length on the Atlantic Ocean, and is called the Mid-Atlantic Ridge. At these zones large amounts carbon dioxide are released. During time of enhanced tectonic activity and sea floor spreading, elevated levels of carbon dioxide emissions may increase the strength of the Earth's natural greenhouse effect.

Different rates of sea-floor spreading can also affect the shape of the seafloor. When tectonic activity is greater, the sea floor is pushed up, leaving a smaller volume to hold the water of the oceans. Consequently, sea levels can rise by several hundred metres, covering large areas of the continents with warm shallow seas. Indeed, during the age of the dinosaurs about 100 million years ago, the sea level was much higher than it is today and the climate was much warmer.

Earth's Orbit

The Milankovitch theory, named after the Yugoslav mathematician who first proposed it, is the astronomical or orbital theory of climate variations. Since these ideas were put forward, much evidence from palaeoclimatic records has been found to support the theory. The original Milankovitch theory identifies three types of variation in the Earth's orbit around the Sun which could act as mechanisms to change the global climate. These include changes in the tilt of the Earth's axis (obliquity), changes in the shape of Earth's orbit (eccentricity) and the shifting of the equinoxes (precession). Each variation has its specific time period, the shortest being 22,000 years and the longest being 96,000 years.

The three orbital variations together affect the total amount of sunlight received by the Earth, and distribution of that sunlight at different latitudes and at different times. With time periods measured in tens of thousands of years, one would expect that changes in climate as a result of orbital variations would occur over similar time periods. Indeed, the Milankovitch theory of climate change has been used to explain the global climate of the last 2 million years, with changes between warmer interglacial periods and colder Ice Ages occurring over a 100,00 year cycle, as predicted by the Milankovitch theory of climate change.

Global Energy Balance

The global energy balance is the balance between incoming energy from the Sun and outgoing heat from the Earth. The global energy balance regulates the state of the Earth's climate, and modifications to it as a result of natural and man-made climate forcing, cause the global climate to change.

Energy released from the Sun as electromagnetic radiation has a temperature of approximately 6000°C. At this temperature, electromagnetic radiation is emitted as shortwave light and ultraviolet energy. Electromagnetic radiation travels across space at the speed of light. When it reaches the Earth's, some is reflected back to space by clouds, some is absorbed by the atmosphere, and some is absorbed at the Earth's surface. The Earth releases a lot of energy it has received from the Sun back to space. However, since the Earth is much cooler than the Sun, its radiating energy is longer wavelength infrared energy or heat. Sometimes, we can indirectly see heat radiation, for example as heat shimmers rising from a tarmac road on a hot sunny day. The energy received by the Earth from the Sun balances the energy lost by the Earth back into space. In this way, the Earth maintains a stable average temperature and therefore a stable climate (although of course differences in climate exist at different locations around the world).

The Earth atmosphere contains a number of greenhouse gases, which affect the Sun-Earth energy balance. The average global temperature is in fact 33°C higher than it should be. Greenhouse gases absorb electromagnetic radiation at some wavelengths but allow radiation at other wavelengths to pass through unimpeded. The atmosphere is mostly transparent in the visible light, but significant blocking (through absorption) of ultraviolet radiation by the ozone layer, and infrared radiation by greenhouse gases, occurs.

The absorption of infrared radiation trying to escape from the Earth back to space is particularly important to the global energy balance. Such energy absorption by the greenhouse gases heats the atmosphere, and so the Earth stores more energy near its surface than it would if there was no atmosphere. The average surface temperature of the moon, about the same distance as the Earth from the Sun, is -18°C. The moon, of course, has no atmosphere. By contrast, the average surface temperature of the Earth is 15°C. This heating effect is called the natural greenhouse effect.

Evidences of Climate Change

During the most recent history, scientists have been able to construct evidence of climate change from collected information on temperature, rainfall and other weather variables from measuring stations all over the world. The oldest time series of climate data is a temperature record from central England beginning in the 17th century. Most instrumental records however, date back only to the 19th century. Despite this relatively short period of data collection, by careful mathematical analysis scientists have been able to demonstrate that the Earth's surface has warmed on average by about 0.6°C during the 20th century. Such warming, it is believed, is related largely to mankind's pollution of the atmosphere by greenhouse gases.

Of course, it is now known that the Earth's climate has changed many times before mankind began altering the global environment, and before we began collecting information about the weather. Knowledge of palaeoclimate changes has come from the reconstruction of indirect or proxy sources of information. For climate changes that have occurred during the last 2,000 years or so, historical records may be used if they contain details of particular weather phenomena that occurred at the time the records were written. Other forms of proxy data come from the study of natural phenomena which are climate-dependent. The growth of tree rings, for example, is dependent on climate, and tree ring sizes and wood densities may be used to reconstruct climates at the time of growth. The oldest surviving trees are about 4,000 years old. Older wood from dead trees has been used to reconstruct past climates back about 10,000 years.

To reconstruct the climate of the last Ice Age, scientists have had to turn to the chemical and physical properties of ice, collected from ice cores drilled in Greenland and Antarctica. As snow is laid down and compacted it stores a record of the current climate. The longest cores recovered are several kilometres long. Such cores maintain a record of climate changes back over 100,000 years to the pervious interglacial warm period. Palaeoclimatologists have also been able to use sea sediments to reconstruct climate during the last Ice Age and many before it, stretching back millions of years. To investigate the oldest climate changes, some stretching back many hundreds of millions of years, geologists have to use clues from rocks, once laid down as sediment at the bottom of oceans, compressed and then uplifted to be exposed on the continents. Such geological evidence has been used to confirm ideas that continental drift and mountain building can change the Earth's climate over hundreds of millions of years.

The Earth's climate responds to changes in the amount of energy stored by the climate system, and in particular when the global energy balance between incoming energy from the Sun and outgoing heat from the Earth is upset. When "climate forcing" upsets the balance of the Earth's climate in this manner, it responds not only to the process forcing the climate to change, but also to feedback effects which can either augment or diminish the original influence.

Today, the Earth is warming up. Global warming, it is believed, is largely the result of man-made pollution of the atmosphere with excess greenhouse gases. Globally, the average surface temperature has increased by about 0.6°C during the last 100 years. In the polar regions however, regional warming has been considerably greater. In some parts of Antarctica and northern Russia, temperatures have increased by about 2°C in only 50 years. Many scientists think that such rapid warming in these ice-covered parts of the world is a consequence of the ice-albedo feedback effect.

Ice, being white, reflects a lot of sunlight and keeps the surface colder for longer. Highly reflective surfaces like ice have high albedos. As an enhanced greenhouse effect warms the surface of the Earth, some of the ice at high latitudes melts, exposing either bare ground or ocean, both of which have lower albedos (or reflectivities) than ice. With a lower albedo, the exposed surfaces reflect less sunlight, with more sunlight being absorbed. This causes a further rise in surface temperature, and in turn a further melting of ice. Here, the climatic response to primary greenhouse heating acts as a secondary climate forcing that augments the initial climate change.

Negative feedback occurs when the response to primary changes acts in the opposite direction to that of the initial climate forcing. Negative feedback reduces the climate response to initial causes of climate change. The formation of clouds in a greenhouse-heated world may cause a negative feedback. A warmer atmosphere will contain more moisture, and consequently more clouds. Clouds reflect a lot of sunlight and may help to reduce the amount of global warming due to man-made greenhouse gas emissions.

Over the longer term, changes in the shape of the Earth's orbit around the Sun are believed to influence natural global climate variations that are evident in the palaeoclimatic records over tens and hundreds of thousand of years. It is clear however, that climate feedback effects have augmented the differences in global climate between the Ice Ages and the warmer interglacial periods, including changes in ocean circulation and atmospheric composition.

Global Warming

Causes of climate change involve any process that can alter the global energy balance between energy coming from the Sun and energy leaving the Earth. There are many natural causes of climate change, but recently we have become concerned with the effect mankind's pollution of the atmosphere may be having on the global climate.

The Earth has a natural greenhouse effect which keeps it much warmer that it would be without an atmosphere. Greenhouse gases in the atmosphere trap infrared heat energy trying to escape back to space. In doing so they raise the temperature of the lower atmosphere and the Earth's surface in contact with it. During the last 200 years, mankind has been releasing substantial quantities of extra greenhouse gases to the atmosphere, through the burning of fossil fuels and deforestation. These extra gases are trapping more heat in the atmosphere, and it is now suspected that the observed warming of the Earth by about 0.6°C since the late 19th century is due to this man-made enhancement of the natural greenhouse effect.

This climatic trend has become known as "global warming", and may be distinguished from historical and prehistorical climate changes that have occurred naturally. The term "global warming" is usually reserved for the observed global climate change during the last 100 to 150 years that is believed to be related to mankind's enhancement of the greenhouse effect.

Greenhouse Effect

The Sun, which is the Earth's only external form of heat, emits solar radiation mainly in the form of shortwave visible and ultraviolet (UV) energy. As this radiation travels toward the Earth, 25% of it is absorbed by the atmosphere and 25% is reflected by the clouds back into space. The remaining radiation travels unimpeded to the Earth and heats its surface. The Earth releases a lot of energy it has received from the Sun back to space. However, the Earth is much cooler than the Sun, so the energy re-emitted from the Earth's surface is much weaker, in the form of invisible longwave infrared (IR) radiation, sometimes called heat.

Greenhouse gases like water vapour, carbon dioxide, methane and nitrous oxide trap the infrared radiation released by the Earth's surface. The atmosphere acts like the glass in a greenhouse, allowing much of the shortwave solar radiation to travel through unimpeded, but trapping a lot of the longwave heat energy trying to escape back to space. This process makes the temperature rise in the atmosphere just as it does in the greenhouse. This is the Earth's natural greenhouse effect and keeps the Earth 33°C warmer than it would be without an atmosphere, at an average 15°C. In contrast, the moon, which has no atmosphere, has an average surface temperature of -18°C.

During the last 200 years mankind has been releasing extra quantities of greenhouse gases which are trapping more heat in the atmosphere. Over the same time period the climate of the Earth has warmed, and many scientists now accept that there is a direct link between the man-made enhancement of the greenhouse effect and global warming.

Greenhouse Gases

Despite their relative scarcity, the so-called greenhouse gases play an important role in the regulation of the Earth's energy balance. Greenhouse gases in the atmosphere trap infrared heat energy trying to escape back to space. In doing so they raise the temperature of the lower atmosphere and the Earth's surface in contact with it. This warming process is called the natural greenhouse effect, but during the last 200 years, mankind's pollution of the atmosphere with extra greenhouse gases has enhanced this natural greenhouse effect that may be contributing to global warming.

Greenhouse gases include any gas in the atmosphere that is capable, as a result of its particular molecular structure, of absorbing infrared radiation or heat. They are called greenhouse gases because they behave like glass in a greenhouse gas, allowing sunlight to pass through but trapping the heat formed and preventing it from escaping, thereby causing a rise in temperature. Natural greenhouse gases include water vapour or moisture, carbon dioxide, methane, nitrous oxide and even ozone, which is more commonly associated with the ozone layer and ultraviolet radiation. The amounts of all these gases in the atmosphere are now being increased as a result of man-made processes, such as fossil fuel burning and deforestation. The atmospheric concentration of carbon dioxide, for example, has increased by 30% since the 18th century, whilst levels of methane have more than doubled. Water vapour, whilst not directly released by man-made processes in substantial quantities, may be increasing as a result of climate feedback effects.

In addition to the man-made increase of naturally occurring greenhouse gases, mankind has released some completely new chemicals into the atmosphere, including the CFCs or chlorofluorocarbons. Although these have now been banned in an attempt to save the ozone layer, they will remain in the atmosphere for at least another 50 years. Although their abundance in the atmosphere is very low, molecule for molecule they can be thousands of times better at absorbing heat than carbon dioxide, and consequently contribute significantly to the enhanced greenhouse effect. Furthermore, their replacements, the HCFCs (hydrochlorofluorocarbons) and HFCs (hydrofluorocarbons), whilst being relatively harmless to the ozone layer, are equally potent greenhouse gases, and at present their phase-out dates are not due for another 20 to 30 years.

Climatic Noise, Variability and Change

This arbitrary but necessary procedure makes possible a formal definition of the terms *climatic noise, variability* and *change:*

— *Climatic noise* may be defined as that part of the variance of climate attributable to short-term weather changes. The chance location of the start and end of the averaging periods with respect to short-term variations may produce small statistical differences between successive periods. Leith confines the term climatic noise to this usage. Others use it more loosely to include, for example, anomalous weather extremes, or all short-term variation.

— *Climatic variability* is best thought of as the manner of variation of the climatic parameters *within* the typical averaging period. Suitable measures of variability include such parameters as the standard deviation of continuous elements like temperature and pressure, and the frequency spectrum of the observed variations (giving such useful measures as return-period, or the probability of successive anomalous months or years). A proper statement of the climate of an averaging period should hence include, in addition to measures of central tendency, suitable estimates of the variability.

— *Climatic change* is said to occur when the differences between successive averaging periods exceed what noise can account for, i.e., when a distinct signal exists that is visible above the noise. A common usage is to refer to short-term changes lasting only a few decades as *climatic fluctuations*, especially if conditions then return to the earlier state. Many authorities reserve the term change for longer-term variation extending over centuries. All such usage depends on the time-scale employed as standard.

Shows estimated annual rainfall over England and Wales from 1766 until 1970. Clearly annual rainfall is a variable element. Within any 30-year period there are likely to be yearly totals as high as 1000 mm, or as low as 800 mm. In the 205 years of thc record, rainfall exceeded 1150 mm five times and fell below 700 mm five times. The time-series shows no significant trend over the period of record, and little autocorrelation. There is obviously high variability, and the mean value obtained for any arbitrary 30-year period is likely to differ from that derived from another choice. In the long perspective, however, the record justifies the generalisation that annual mean rainfall over England and Wales, though very variable from year to year, has not changed greatly over the period.

Shows a similar estimate for spatially averaged rainfall over monsoon India. Again there is no significant trend over the entire period 1865 1970. If, however, one inspects the curve starting in 1900 and ending in 1960, a distinct upward trend can be seen. This was a vital climatic fluctuation that had a real bearing on India's capacity to increase food production in pace with her population explosion. The abrupt downward sag in the mid-1960s was an unwelcome surprise. The record shows a tendency for wet or dry years to follow one another, and also a small but statistically significant 2½- to 3-year periodicity. Here again, however, the most central fact is that no lasting change in mean annual rainfall can be demonstrated. It is obviously dangerous to base long-term conclusions on short-term records or to assume that trends will continue indefinitely.

These two records illustrate another general principle. Any trends that may be present in a climatic time-series, and especially any persistent climatic changes, are usually very small by comparison with short-term variations. The latter obscure the signal corresponding to the change, just as radio noise may obscure the desired radio signal. It is not yet easy, for example, to detect the signal corresponding to the expected effects of carbon dioxide increase on the trend of air temperature, because short-term variations dominate the record. Public recognition of these facts is very partial. A sequence of wet or dry years, as displayed in shows, often raises false hopes or fears, and leads to speculation about lasting change. Institutional memory of severe extremes tends to fade quickly, but the public at large retains a recollection of events like cold winters or dry summers.

Climate Change Factors

Climate changes reflect variations within the Earth's environment, natural processes going on around it, and the impact of human activity. The external factors that can shape climate are often called climate forcings and include such processes as variations in solar radiation, the Earth's orbit, and greenhouse gas concentrations.

Variations within the Earth's Climate

Weather, in and of itself, is a chaotic non-linear dynamical system, but in many cases, it is observed that the climate is fairly stable and predictable. This includes the average temperature, amount of precipitation, days of sunlight, and many other variables that might be measured at any given site. However, there are also changes within the Earth's environment that can affect the climate.

Glaciation

Glaciers are recognised as one of the most sensitive indicators of climate change, advancing substantially during climate cooling and retreating during climate warming on moderate time scales. Glaciers grow and collapse, both contributing to natural variability and greatly amplifying external forces. For the last century, however, glaciers have been unable to regenerate enough ice during the winters to make up for the ice lost during the summer months.

The most important climate processes of the last several million years are the glacial and interglacial cycles of the present ice age. Though shaped by orbital variations, the internal responses involving continental ice sheets and 130 m sea-level change certainly played a key role in deciding what climate response would be observed in most regions. Other changes, including Heinrich events, Dansgaard–Oeschger events and the Younger Dryas show the potential for glacial variations to influence climate even in the absence of specific orbital changes.

Ocean Variability

On the scale of mere decades, climate changes can also result from changes within the ocean/atmosphere systems. Many climate states, most obviously El Nino Southern oscillation, but also including the Pacific decadal oscillation, the North Atlantic oscillation, and the Arctic oscillation, have been recognised as modes within the climate system, owing their existence at least in part to different ways that heat can be stored in the oceans and move between different reservoirs. On longer time scales, ocean processes such as thermohaline circulation play a key role in redistributing heat, and could, if changed, dramatically impact climate.

Memory of Climate

More generally, most forms of internal variability in the climate system can be recognised as a form of hysteresis, meaning that the current state of climate reflects not only the inputs, but also the history of how it got there. For example, a decade of dry conditions may cause lakes to shrink, plains to dry up and deserts to expand. In turn, these conditions may lead to less rainfall in the following years. In short, climate change can be a self-perpetuating process because different aspects of the environment respond at different rates and in different ways to the fluctuations that inevitably occur.

Non-climate Factors Driving Climate Change

Greenhouse gases

Current studies indicate that radiative forcing by greenhouse gases is the primary cause of global warming. Greenhouse gases are also important in understanding Earth's climate history. According to these studies, the greenhouse effect, which is the warming produced as greenhouse gases trap heat, plays a key role in regulating Earth's temperature.

Over the last 600 million years, carbon dioxide concentrations have varied from perhaps >5000 ppm to less than 200 ppm, due primarily to the impact of geological processes and biological innovations. Curiously, it has been argued that variations in greenhouse gas concentrations over tens of millions of years have not been well correlated to climate change, with plate tectonics perhaps playing a more dominant role. However, there are several examples of rapid changes in the concentrations of greenhouse gases in the Earth's atmosphere that do appear to correlate to strong warming, including the Paleocene–Eocene thermal maximum, the Permian–Triassic extinction event, and the end of the Varangian snowball earth event. During the modern era, rising carbon dioxide levels are implicated as the primary cause of global warming since 1950.

Plate tectonics

On the longest time scales, plate tectonics will reposition continents, shape oceans, build and tear down mountains and generally serve to define the stage upon which climate exists. More recently, plate motions have been implicated in the intensification of the present ice age when, approximately 3 million years ago, the North and South American plates collided to form the Isthmus of Panama and shut off direct mixing between the Atlantic and Pacific Oceans.

Solar variation

The sun is the ultimate source of essentially all heat in the climate system. The energy output of the sun, which is converted to heat at the Earth's surface, is an integral part of shaping the Earth's climate. On the longest time scales, the sun itself is getting brighter with higher energy output; as it continues its main sequence, this slow change or evolution affects the Earth's atmosphere. Early in Earth's history, it is thought to have been too cold to support liquid water at the Earth's surface, leading to what is known as the Faint young sun paradox.

On more modern time scales, there are also a variety of forms of solar variation, including the 11-year solar cycle and longer-term modulations. However, the 11-year sunspot cycle does not manifest itself clearly in the climatological data. Solar intensity variations are considered to have been influential in triggering the Little Ice Age, and for some of the warming observed from 1900 to 1950. The cyclical nature of the sun's energy output is not yet fully understood; it differs from the very slow change that is occurring to the sun as it ages and evolves.

Orbital variations

In their impact on climate, orbital variations are in some sense an extension of solar variability, because slight variations in the Earth's orbit lead to changes in the distribution and abundance of sunlight reaching the Earth's surface. Such orbital variations, known as Milankovitch cycles, are a highly predictable consequence of basic physics due to the mutual interactions of the Earth, its moon, and the other planets. These variations are considered the driving factors underlying the glacial and interglacial cycles of the present ice age. Subtler variations are also present, such as the repeated advance and retreat of the Sahara desert in response to orbital precession.

Volcanism

A single eruption of the kind that occurs several times per century can impact climate, causing cooling for a period of a few years. For example, the eruption of Mount Pinatubo in 1991 is barely visible on the global temperature profile. Huge eruptions, known as large igneous provinces, occur only a few times every hundred million years, but can reshape climate for millions of years and cause mass extinctions. Initially, scientists thought that the dust emitted into the atmosphere from large volcanic eruptions was responsible for the cooling by partially blocking the transmission of solar radiation to the Earth's surface. However, measurements indicate that most of the dust thrown in the atmosphere returns to the Earth's surface within six months.

Human influences on climate

Anthropogenic factors are acts by humans that change the environment and influence climate. The biggest factor of present concern is the increase in CO_2 levels due to emissions from fossil fuel combustion, followed by aerosols (particulate matter in the atmosphere) which exerts a cooling effect. Other factors, including land use, ozone depletion, animal agriculture and deforestation also impact climate.

Fossil fuels

Beginning with the industrial revolution in the 1850s and accelerating ever since, the human consumption of fossil fuels has elevated CO_2 levels from a concentration of ~280 ppm to more than 370 ppm today. These increases are projected to reach more than 560 ppm before the end of the 21st century. Along with rising methane levels, these changes are anticipated to cause an increase of 1.4–5.6 °C between 1990 and 2100.

Aerosols

Anthropogenic aerosols, particularly sulphate aerosols from fossil fuel combustion, are believed to exert a cooling influence. This, together with natural variability, is believed to account for the relative "plateau" in the graph of 20th century temperatures in the middle of the century.

Land use

Prior to widespread fossil fuel use, humanity's largest impact on local climate is likely to have resulted from land use. Irrigation, deforestation, and agriculture fundamentally change the environment. For example, they change the amount of water going into and out of a given locale. They also may change the local albedo by influencing the ground cover and altering the amount of sunlight that is absorbed. For example, there is evidence to suggest that the climate of Greece and other Mediterranean countries was permanently changed by widespread deforestation between 700 BC and 0 BC, with the result that the modern climate in the region is significantly hotter and drier, and the species of trees that were used for ship-building in the ancient world can no longer be found in the area.

A controversial hypothesis by William Ruddiman suggests that the rise of agriculture and the accompanying deforestation led to the increases in carbon dioxide and methane during the period 5000–8000 years ago. These increases, which reversed previous declines, may have been responsible for delaying the onset of the next glacial period, according to Ruddimann's hypothesis.

Animal agriculture

According to a 2006 United Nations report, animal agriculture is responsible for 18% of the world's greenhouse gas emissions as measured in CO_2 equivalents. By comparison, all transportation emits 13.5% of the CO_2. In addition to increased CO_2 emissions, animal agriculture produces 65% percent of human-related nitrous oxide (which has 296 times the global warming potential of CO2) and 37% of all human-induced methane.

Interplay of Factors

If a certain forcing acts to change the climate, then there may be mechanisms that act to amplify or reduce the effects. These are called positive and negative feedbacks. As far as is known, the climate system is generally stable with respect to these feedbacks: positive feedbacks do not "run away". Part of the reason for this is the existence of a powerful negative feedback between temperature and emitted radiation: radiation increases as the fourth power of absolute temperature. However, a number of important positive feedbacks do exist. The glacial and interglacial cycles of the present ice age provide an important example. It is believed that orbital variations provide the timing for the growth and retreat of ice sheets. However, the ice sheets themselves reflect sunlight back into space and hence promote cooling and their own growth, known as the ice-albedo feedback. Further, falling sea levels and expanding ice decrease plant growth and indirectly lead to declines in carbon dioxide and methane. This leads to further cooling.

Similarly, rising temperatures caused, for example, by anthropogenic emissions of greenhouse gases could lead to retreating snow lines, revealing darker ground underneath, and consequently result in more absorption of sunlight. Water vapor, methane, and carbon dioxide can also act as significant positive feedbacks, their levels rising in response to a warming trend, thereby accelerating that trend. Water vapor acts strictly as a feedback, unlike the other major greenhouse gases, which can also act as forcings.More complex feedbacks involve the possibility of changing circulation patterns in the ocean or atmosphere. For example, a significant concern in the modern case is that melting glacial ice from Greenland will interfere with sinking waters in the North Atlantic and inhibit thermohaline circulation. This could affect the Gulf Stream and the distribution of heat to Europe and the east coast of the United States.

Other potential feedbacks are not well understood and may either inhibit or promote warming. For example, it is unclear whether rising temperatures promote or inhibit vegetative growth, which could in turn draw down either more or less carbon dioxide. Similarly, increasing temperatures may lead to either more or less cloud cover. Since on balance cloud cover has a strong cooling effect, any change to the abundance of clouds also impacts climate. In all, it seems likely that overall climate feedbacks are negative, as systems with overall positive feedback are highly unstable.

Current Status of Climate

Scientists use "Indicator time series" that represent the many aspects of climate and ecosystem status. The time history provides an historical context. Current status of the climate is also monitored with climate indices.

Evidence for Climatic Change

Evidence for climatic change is taken from a variety of sources that can be used to reconstruct past climates. Most of the evidence is indirect—climatic changes are inferred from changes in indicators that reflect climate, such as vegetation, dendrochronology, ice cores, sea level change, and glacial retreat.

Pollen Analysis

Species have particular climatic requirements that influence their geographical distributions. Each plant species has a distinctively shaped pollen grain, and if these fall into oxygen-free environments, such as peat bogs, they resist decay. Changes in the pollen found in different levels of the bog indicate, by implication, changes in climate. One limitation of this method is the fact that pollen can be transported considerable distances by wind or sometimes by wildlife.

Coleoptera

Remains of Coleoptera beetles are common in freshwater and land sediments. Different species of this beetle tend to be found under different climatic conditions. Knowledge of the present climatic range of the different species, and the age of the sediments in which remains are found, allows past climatic conditions to be worked out.

Glacial Geology

Advancing glaciers leave behind moraines and other features that often have datable material in them, recording the time when a glacier advanced and deposited a feature. Similarly, the lack of glacier cover can be identified by the presence of datable soil or volcanic tephra horizons. Glaciers are considered one of the most sensitive climate indicators by the IPCC, and their recent observed variations provide a global signal of climate change.

Historical Records

Historical records include cave paintings, depth of grave digging in Greenland, diaries, documentary evidence of events and evidence of areas of vine cultivation. Since 1873, daily weather reports have been documented, and the Royal Society has encouraged the collection of data since the seventeenth century. Parish records are often a good source of climate data.

Public Perception of Climate Change

There is a widespread belief that climate is unstable, and may be undergoing change. This is not a new impression. Literary sources show that speculation about climatic change is a venerable human tradition. Yet the changes in world climate during mankind's period of literacy have been small by comparison with those of Pleistocene times.

Public speculation about such issues has been hampered by a failure to distinguish between local and global factors in the control of climate. Major land use changes, which are highly visible, have significant impacts on local radiative balances, and to some extent on temperature. Their visibility has led to speculation that still greater changes may alter local or even global climate. Actually, however, global processes are on such a scale that they are not easily affected. It is the general circulation of the atmosphere and the planetary scale of energy exchanges that ultimately control non-seasonal temperature and humidity variations even at local sites, via the mechanism of advection. Precipitation, too, comes largely from systems controlled by global rather than local processes. Nevertheless popular apprehensions persist, perhaps with some reason, that human manipulation of the earth must lead to climatic change, perhaps of a hostile kind.

Present-day speculation seems to spring from certain recognisable sources, such as popular memory of climatic extremes, widespread media comment on the subject, gloomy statements by professional climatologists, and real fluctuations of temperature during the past century. In particular, certain specific changes seem to have impressed political opinion:

— The decrease in northern hemisphere surface air temperature between 1938 and 1965, which was of an order of 0.5 °C. This decrease has sometimes been announced by the media as the onset of a new glacial period. In actual fact the decrease has reversed since 1965, and is in any case strongly regional, affecting some areas much more than others. A parallel downward trend of sea-surface temperature has been documented. Neither sea nor air temperature shows the same tendency in the southern hemisphere; in fact a reverse movement has been claimed.

— The major west-European drought of 1975-76, which affected densely populated countries with vital agricultural areas. The drought was the most severe in more than two centuries of record in some districts, where rainfall over the crop season was as little as 40 percent of normal. The drought was short by comparison with, for example, the Sahelian drought, but exposed the vulnerability of water-supply systems and crops to rainfall anomalies.

— Severe winters in eastern North America during the middle 1970s, which badly stressed the energy delivery systems of the world's most extensively industrialised region. In some areas temperatures were more than 5°C below normal over the entire 3-month season.

— Disturbances during the 1970s of the world's food system due to quite small reductions of cereal supply caused by climatic anomalies occurring at a time of increasing use of cereal as livestock feed. The crop deficiencies in the Soviet Union in 1972, 1974, 1975, 1979, 1980 and 1981 were particularly significant because they led to large Soviet grain purchases on the world market, and to dramatic fluctuations of price.

— The Sahelian drought, popularly supposed to have occurred in the years 1968-73, but actually a gradual desiccation of much of north and east sub-Saharan Africa that began in 1955 60, an effect that has continued until the present. Because it imperilled many fledgling African states this desiccation achieved high visibility, leading in 1977 to the UN Conference on Desertification, the first world political conference about a climatically related stress.

At the World Climate Conference in Geneva, no evidence was brought forward from any part of the world to suggest that present climate is in the process of change, although it displays the high variability normal to its behavior. The most serious impacts on society have been brought about by short and drastic anomalies of the sort listed above. The time-series of emphasise, for example, that in drought-prone monsoon India the rainfall anomalies are chiefly short but large departures from a basically stable regime. The point is often made, however, that lasting climatic change, if it occurs, will mimic short fluctuations. We may not recognise the change until it has been in progress for decades. A decrease in mean rainfall, for example, will probably be accompanied by a tendency towards more frequent dry years. It will be extremely difficult to recognise climatic change in time to do anything about it.

Just as there is little evidence for significant changes of mean temperature, rainfall or most other measured elements, there is also little evidence that their variability is changing significantly. Many investigations show that variability does in fact fluctuate in magnitude and character. Thompson, Haigh, McQuigg and others have suggested that reduced rainfall variation in the US Midwest in the late 1950s and 1960s was responsible for part of the observed increase in corn yields in that period. Haigh's analysis does indeed demonstrate that an upward trend of mean July precipitation and a downward trend of mean August temperature affected the state of Illinois between 1930 and 1976, and that variability of rainfall was markedly lower in the middle of that period. But there is no consensus that overall variability of climate has recently changed in any large region of the earth. Such variability is related in most regions to persistent anomalies of the general circulation of the atmosphere. The analysis by Van Loon and Wiliams cited above shows that such anomalies can have effects on temperature over 15-year averaging periods.

Clearly the impact of climatic variability on all time-scales will depend on the ability to predict individual variations with sufficient skill to make precautionary measures possible. Unfortunately there is a professional consensus that such predictions are as yet rarely possible, even under favorable circumstances. Present day short-term numerical weather prediction models are effective only up to 4- or 5-day periods, except in areas where weather changes are slow. Increases in computer capability are slowly improving this situation, but the cost of even a day's extension in reliable forecasts is very high. Moreover the nature of weather systems makes it doubtful *in principle* whether such an approach can be pushed very far. Tentative foreshadowing on the monthly or seasonal scale can be based on analogue methods, on assumed sun-spot or orbital variations, and on assumed relationships with sea-surface temperature anomalies. Some of these attempts achieve better than chance results, but none has yet reached the point where management decisions can be firmly based on the outcome.

The search for such predictability is the major objective of the research component of the World Climate Programme. It is premature to prejudge the results of the large-scale effort now being launched, but the following generalisations appear sound:

— Attempts to model future climatic variation resulting from the build-up of carbon dioxide and other infra-red absorbers yield predictions of trend, on the order of 0.1 °C a-1 or less. Though this is not negligible, it has relevance mainly for long-range planning of such things as plant breeding, navigation in ice-infested waters, and dam construction. The models do not predict future variability with any assurance.

— The best hope for seasonal or interannual prediction may well rest on two existing bodies of active research: sea-surface temperature anomaly to atmosphere relationships, and possible large-scale oscillations in atmosphere ocean behavior. Among the latter the most promising is the so-called *southern oscillation.*

The southern oscillation has been known, though not understood, for 60 years, having served for many years as the basis for statistical forecasting of Indian monsoon rainfall. It consists of a large longitudinal quasi-periodic oscillation of atmospheric pressure, representing gigantic transfers of mass in east-west directions between the South Atlantic and South Pacific at one pole, and a broad area between the Red Sea, East Africa and Australasia at the other. Pressure is negatively correlated between these two large areas, but unfortunately the period of oscillation varies. The average period is 38 months, but individual oscillations have varied between 2 and 10 years.

The southern oscillation is highly correlated with sea temperature behavior over the equatorial Pacific. Shortly after the onset of the atmospheric event the normal cold upwelling off the Peruvian coast is weakened, and much warmer water appears nearshore, which badly affects the highly productive marine ecosystem the so-called *El Nino* event. In the typical case the warmest water shifts westwards along the equatorial belt of the Pacific, and within 18 months cold upwelling is re-established off Peru. The dramatic effect of these events on fisheries has given them high visibility in recent decades.

This entire phenomenon is a vast longitudinal perturbation of atmosphere and ocean that has hitherto eluded adequate theoretical understanding. The World Climate Research Programme will give such understanding very high priority. The phenomenon seems to be meaningfully correlated with climatic variation outside the directly affected regions for example with air pressure and temperature over North America. Unfortunately the regularities are in most cases insufficient for successful forecasting. There is first the irregular period and erratic behavior of the southern oscillation events, and second the weakness of the observed correlations. Only theoretical understanding and effective modelling can serve to reduce these uncertainties and make reliable prediction possible.

Predictability may well come in the future, for at least some of these phenomena. Meanwhile, there is abundant and conclusive evidence that variability of the main climatic elements is a normal and necessary part of nearly all climates. This variability extends to significant fluctuations over several decades. But most long climatic time-series show an underlying stability, suggesting that the variability is distributed around a fundamentally unchanging normal state. Investigations have failed to show that changes in variability tend to occur as mean temperatures rise and fall. They have also failed to confirm truly periodic anomalies. The recurrence of drought somewhere on the Great Plains of the United States in the 1890s, 1910s, 1930s, 1950s, and, to a limited degree, the 1970s, for example, is not seen as truly periodic. Such quasi-periodic happenings are probably misleading as to future climate, as evidenced in the unrealised expectation of widespread drought on the Great Plains in the 1970s.

Impacts of Climate Change

Since the end of the 19th century the earth's average surface temperature has increased by 0.3-0.6 °C. Over the last 40 years, the rise has been 0.2-0.3 °C. Recent years have been the warmest since 1860, the year when regular instrumental records became available. In 1995, a group of 2000 leading scientists got together under the IPCC and concluded that global warming is real, serious, and accelerating. They predicted that over the next 100 years, the earth's average temperature could further rise by 1.4-5.8 °C. The magnitude of this predicted warming may seem negligible, but its rate is faster than any seen in the last 10, 000 years. Some important aspects of our lives can be affected through changes in weather patterns and some of these are discussed here.

Agriculture

The steadily-increasing human population has led to a rise in the demand for food. As more land comes under agricultural cultivation there will be more pressure on natural ecosystems. Climate change will affect agricultural yield directly because of alterations in temperature and rainfall, and indirectly through changes in soil quality, pests, and diseases. In particular, the yield of cereals is expected to decline in India, Africa, and the Middle East. As the temperature rises conditions will become more favourable for pests such as grasshoppers to complete a number of reproduction cycles thereby increasing their population. In the higher latitudes (in the northern countries) agriculture will benefit with the rise in temperature as the winter season will be shorter and the growing seasons longer. This will also mean that pests that will move towards the higher latitudes as the temperatures rise. Extreme weather conditions such as high temperature, heavy rainfall, floods, droughts, etc. will also affect crop production.

Weather

A warmer climate will change rainfall and snowfall patterns, lead to increased droughts and floods, cause melting of glaciers and polar ice sheets, and result in accelerated sea- level rise. Rising warmth will lead to an increase in the level of evaporation of surface water; the air will also expand and this will increase its capacity to hold moisture. This, in turn, will affect water resources, forests, and other natural ecological systems, agriculture, power generation, infrastructure, tourism, and human health. An increase in the number of cyclones and hurricanes over the last few years has been attributed to changes in temperature.

Sea Level Rise

Coastal areas and small islands are among the most densely-populated parts of the world. They are also the most threatened because of rises in sea level that global warming may cause. The heating of oceans, and melting of glaciers and polar ice sheets, is predicted to raise the average sea level by about half a metre over the next century. Sea-level rise could have a number of physical impacts on coastal areas, including loss of land due to inundation and erosion, increased flooding, and salt-water intrusion. These could adversely affect coastal agriculture, tourism, freshwater resources, fisheries and aquaculture, human settlements, and health. Rising sea levels threaten the survival of many low-lying island nations, such as the Maldives and Marshall Islands.

There could be large decreases in the availability of water in many rivers because of rainfall and snow. The volume in others would increase due to glaciers melting, for example, the rivers originating in the Himalayas. Shifts in water availability could also affect hydropower generation, and industries such as paper, pharmaceutical, and chemical manufacturing, that use large quantities of water. Buildings and other infrastructure would be vulnerable to any increase in the frequency of storms and other extreme events, which could also disrupt transport routes.

Forests and Wildlife

Ecosystems sustain the earth's entire storehouse of species and genetic diversity. Plants and animals in the natural environment are very sensitive to changes in climate. The ecosystems that are most likely to be affected by this change are the ones in the higher latitudes, the tundra forests. Polar regions will feel the impact of warming more than others. Interiors of continents will experience more warming than the coastal regions.

National parks are supposed to provide a sanctuary to wildlife from the ravages of humankind on nature. But no park boundary or conservation law can protect an ecosystem from climate change. A recent report by the WWF (The World Wide Fund for Nature) states that this invisible killer has entered the most cherished natural areas. The giant pandas of Wolong in China, the grizzly bears of America's Yellowstone National Park, and the tigers in Kanha National Park in India are some of the animals at risk from global warming. Mountain parks have been identified as being especially at risk from the environmental destruction caused by climate change. Species that live in the higher alpine zones, are forced to move higher up to find a suitable habitat thus reducing the area in which they can live. If the rate of climate change continues to accelerate, then the extinction of some mountain plants and animals is certain. Migratory birds fly from the cold northern parts of the world to the warmer south. Factors such as the weather and food sources along the route are very important for the successful completion of their journey. Changes in climate may bring about a shift in their feeding points and disruptions to their flight patterns.

Marine Life

Corals are known as the tropical forests of the oceans and sustain diverse life forms. As ocean waters in the tropics become warmer, the damage to coral reefs seems to be increasing. These corals are very sensitive to changes in water temperature, which causes bleaching. Large stretches of the Great Barrier Reef in Australia have been damaged by bleaching. There is still a great deal that we do not understand about our climate, and about how our activities will change it. But one thing can be said loud and clear: if we wait to get answers to these questions, it will probably be too late!

Health Impacts

Climate change is a major problem caused by the increase of human activities leading to several direct and indirect impacts on health. The combustion of fossil fuels, increasing number of industries, and large-scale deforestation are some of the causes for the accumulation of GHGs (greenhouse gases) in the atmosphere. According to the IPCC, an increase in carbon dioxide and other GHGs, like methane, ozone, nitrous oxide, and chlorofluorocarbons, in the atmosphere is expected to increase the average global temperature by 1.5°C to 4.5°C. This in turn will lead to changes in rainfall and snowfall, more intense or frequent droughts, floods, and storms, as well as a rise in sea level. These climatic changes will have wide-ranging harmful effects including increase in heat-related mortality, dehydration, spread of infectious diseases, malnutrition, and damage to public health infrastructure. Thus we should take appropriate measures to stop this climate change.

Direct impacts

The weather has a direct impact on our health. If the overall climate becomes warmer, there will be an increase in health problems. It is anticipated that there will be an increase in the number of deaths due to greater frequency and severity of heat waves and other extreme weather events. The elderly, the very young and those suffering from respiratory and cardiovascular disorders will probably be affected by such weather extremes as they have lesser coping capacity. An extreme rise in the temperature will affect people living in the urban areas more than those in the rural areas. This is due to the 'heat islands' that develop here owing to the presence of concrete constructions, paved and tarred roads. Higher temperatures in the cities would lead to an increase in the ground-level concentration of ozone thereby increasing air pollution problems.

Indirect impacts

Indirectly, changes in weather pattern, can lead to ecological disturbances, changes in food production levels, increase in the distribution of malaria, and other vector-borne diseases. Fluctuation in the climate especially in the temperature, precipitation, and humidity can influence biological organisms and the processes linked to the spread of infectious diseases.

Higher temperature will cause the sea levels to rise that could lead to erosion and damage to important ecosystems such as wetlands and coral reefs. Direct impact of this rise would include deaths and injury caused by intense flooding. Temperature rise would indirectly result in geohydrological changes along the coastline such as saltwater intrusion into the groundwater and the wetlands, coral reef destruction, and damage to the drainage in the low-lying areas. Climate change could increase air pollution levels by accelerating the atmospheric chemical reactions that produce photochemical oxidants due to a rise in the temperature.

Diseases

The GHGs have been responsible for the depletion of stratospheric ozone, which protects the earth from the harmful direct rays of the sun. Depletion of stratospheric ozone results in higher exposure to ultra violet rays of the sun, leading to an increase in the incidents of skin cancer in light skinned people. It could also lead to an increase in the number of people suffering from eye diseases such as cataract. It is also thought to cause suppression of the immune system. Due to global warming there will be an increase in the areas of habitat of disease-spreading insects such as the mosquito, causing an increase in the transmission of infection through these carriers.

Potential effects on health due to sea level rise include:

- death and injury due to flooding;
- reduced availability of fresh water due to saltwater intrusion;
- contamination of water supply through pollutants from submerged waste dumps;
- change in the distribution of disease-spreading insects;
- effect on the nutrition due to a loss in agriculture land and changes in fish catch; and
- health impacts associated with population displacement.

Preventive measures

Reduction in the use of non-renewable sources of energy and increased use of renewable sources will undoubtedly decrease the emission of GHGs substantially. This decrease in the GHGs will have a positive affect on the health and well being of the people. Furthermore, switching to cleaner fuels and energy-efficient technologies will reduce local pollutants and therefore, have an added beneficial impact on health.

Future of Climatic Change

If present and recent past climates show little sign of change, why is there so much concern about the possibility of climatic disruptions of the economy? Two answers are clearly possible, both of which may be correct. One is that human society is becoming more vulnerable to climatic impact as population grows, as the demand for food increases, and as other pressures on the food and energy systems increase. The other answer says that processes are now at work that will upset the stability of present climate. Conceivably we may be able to predict the resulting changes, so that they will not catch the world unawares.

The second view is widely held among climatologists. Since the major disruptions of the international grain trade following the crop year 1972-73, much research has been conducted to test the stability of present climate. This has consisted of one-, two- and three-dimensional modelling exercises representing the present climate as a function of present-day controls, and examination of the models' sensitivity to hypothetical changes in these controls. The sensitivities tested include the dependence of world temperature and precipitation on:

— variations in solar energy input;
— variations in atmospheric optical properties due to altered carbon dioxide concentration, the presence of pollutants such as the chlorofluoromethanes, and higher levels of particle load due to human activity or vulcanism; and
— changes in air ocean interactions, or in the extent of glaciation.

On the time-scales of importance here interannual or interdecadal only the second kind of disturbance seems likely to be effective. In any case these turn out to be similar to those that might follow a very small change in the solar constant. There have been many attempts to establish air-sea interactions by empirical methods, and some of these have been effective. But these appear to relate most closely to fairly brief fluctuations of climate. Specific patterns of sea temperature anomaly are found to be connected with air temperature or precipitation anomalies elsewhere, occasionally with some useful time lag that may make prediction possible. As such the sensitivities of type 3 above relate more to the internal variability of existing climate than to lasting climatic change. For the latter to occur it is now fairly well agreed that changes in external forcing are necessary.

CO_2 Levels

The change in forcing seen most likely to occur is that to be expected from the build-up of carbon dioxide (CO_2) in the atmosphere. Shows the observed change of CO_2 concentration in the atmosphere since serious monitoring began in 1957 58. The observational network is now worldwide, and it is clear that the build-up is a planetary process. The 1983 concentration was near 342 ppmv (parts per million by volume), equivalent to 724 x 109 metric tons of elemental carbon in the atmospheric reservoir. The rate of increase varies from 0.5 to 2.2 ppmv a^{-1}, averaging recently near 1.2 ppmv a^{-1}. In all, a little under 3 x 109 metric tons of carbon are being added to the atmospheric reservoir each year. The source is probably chiefly fossil fuel consumption, currently a little below 6 x 109 metric tons a^{-1}, which implies an atmospheric retention rate of a little over 50 percent. The balance is disappearing into some other reservoir, probably the surface and intermediate waters of the ocean. Atmospheric mixing is so efficient that both additions to and subtractions from the atmospheric store can be thought of as global processes; the CO_2 build-up is not, like acid rain, a focal or regional phenomenon.

`Forcing' is useful jargon for an influence or factor whose behavior is external to the model itself: it `forces' the model, but is not thereby affected. Such a rapid increase of CO_2 necessarily changes the optical properties of the atmosphere. CO_2 is largely transparent to solar radiation, but it strongly absorbs the return long-wave terrestrial radiation near 15 μm, a wavelength at which water vapor is ineffective as an absorber. The effect of CO_2 build-up is thus to increase the resistance to the upward radiative transfer of heat in the atmosphere. To achieve the necessary global balance between incoming solar and outgoing terrestrial radiation the earth's surface must hence become a little warmer, and the stratosphere a little cooler. This is the greenhouse effect so extensively discussed in the popular press.

It seems clear that the build-up will continue, since its presumed causes fossil fuel consumption, probably reinforced by the effects of forest clearance and soil impoverishment are unlikely to abate. Estimates by Häfele and his associates at the International Institute for Applied Systems Analysis suggest for the year 2030 a total world power consumption at the rate of 18 36 x 10^{12} Watt (terawatt = TW). Rotty and Marland give a similar estimate of 27 TW in 2025, a allows for economic restraints to rapid expansion. If correct, this represents a growth from present consumption (8.2 TW) of only 2 percent per annum. Recent data show that between 1950 and 1973 fossil fuel consumption grew at 4.58 percent per annum; since then it has risen at only 2.25 percent.

Rotty and Marland estimate that about three-quarters of the 2025 power demand will be met by fossil fuel burning a level of CO_2 release almost three times the present value. Allowing for a 25 percent error on either side, they estimate a doubling of present CO_2 levels (i.e., from 342 to 684, ppmv) by 2060 at the earliest, and most probably after 2075. An expert World Climate Programme panel chaired by Bolin concluded that in 2025 the atmospheric concentration will actually be in the range 410 to 490 ppmv, with 450 ppmv the most likely value. This is clearly far below earlier estimates, some of which visualised a doubling of present values by 2020 or 2030. Nevertheless it is vital to remember that the use of fossil fuels may proceed more rapidly than these estimates suggest, and the build-up of CO_2 with it.

Mathematical modelling of the potential climatic impact of such increases has been undertaken by several groups. Most expert opinion favors the use of *general atmospheric circulation models* (GCMs), especially those that interact effectively with the ocean, and allow for a realistic geography. Such models predict, for a doubling of CO_2 concentration, a rise of global mean air temperature of 2.0-3.5 °C, with slightly lower values in the tropics and much higher values in north polar latitudes (of order 5-8 °C). Much of the warming comes, in fact, from the positive feedback introduced by the water vapor added to the air as a result of the CO_2 forcing. Two syntheses suggest that the best estimate is that changes in global temperatures on the order 3° ± 1.5 °C will occur for doubled CO_2.

If this increase of 3 °C does follow a doubling of CO_2, it will occur at some date after 2060, if the most rapid Rotty Marland estimate of build-up occurs. This implies a mean global temperature increase averaging only about 0.04 °C a^{-1}, which will be so slow as to be masked by the short-term variability of the present climate. Moreover the heat capacity of the oceans may delay the expected response by some years, perhaps two decades. We must thus face the probability that the impending change of climate will remain unidentified until it is well advanced.

In an attempt to test this possibility Madden and Ramanathan decided to test the temperature record near 60 °N, where, as indicated above, the CO_2 warming should be considerably larger than for the globe as a whole. Treating all other variations of temperature as noise, and confining themselves to summer, when noise is lowest, they showed that a 20-year average should display the CO_2 signal above the noise. In fact, however, the period 1956 75 was not warmer than the period 1906 25; if anything it was marginally colder. Hence they found it impossible to demonstrate that the CO_2 effect was in progress.

Wigley and Jones made a detailed study of signal-to-noise ratios by comparing their own estimates of observed temperatures with model predictions of CO_2 increase by Manabe and Stouffer. Highest values of the ratio occur in summer, and in annual mean surface temperatures averaged over the northern hemisphere or over midlatitudes. The effects Of CO_2...'may not be detectable,' they wrote, `until around the turn of the century. By this time, atmospheric CO_2 concentration will probably have become sufficiently high... that a climatic change significantly larger than any which has occurred in the past century could be unavoidable.'

On the other hand, it is widely held in the Soviet Union that the signal is already visible. There have also been model simulations incorporating the assumed CO_2 warming,with other influences that account quite well for climatic variation over the past century. To assess such contradictions, what is needed is agreement on the statistical tests to be applied and on rigorous monitoring of the actual unfolding of events in the next few decades. Given the high importance of the outcome, major effort needs to be expended on such monitoring.

Assuming the CO_2 effect to be real, and to be roughly of the character foreshadowed in this section, we can speculate as to further climatic consequences. The major effect may be on the hydrologic cycle. There have been two complementary approaches to this question:

— Some of the GCM experiments show a small overall increase in world precipitation for a doubled or quadrupled CO_2 concentration. A marked decrease in available soil water is foreseen, however, for a midlatitude belt, chiefly because of increased evapotranspiration. The resolution of the models does not allow any statement as to which longitudes may be affected. World climate is always longitudinally differentiated.

— In an attempt to provide such longitudinal detail, Kellogg and Schware compiled a world map of possible soil moisture patterns on a warmer earth. The map synthesised (a) geological experience of distant past epochs, notably the mid-Holocene Altithermal, (b) recent experience of hot and cold years in the northern hemisphere, and (c) the modeling exercises referred to above. The map showed drier conditions over interior North America, Siberia, southeastern South America, the tropical rainforest areas and the polar zones. Wetter conditions occurred in most subtropical countries and west coast temperate areas.

Such foreshadowings show the vital importance in climatic impact studies of the CO_2 effect, acting indirectly via the effect of differential atmospheric heating on the hydrologic cycle. So far the model estimates have withstood a number of objections raised against them.

Rising CO_2 concentrations may also have an effect on polar snow and ice distribution. Snow and ice surfaces have a high albedo. If general warming causes their equatorward limit to recede poleward, a significant increase in the absorption of solar radiation occurs, which intensifies the heating. This albedo feedback mechanism is built into the models described above, and accounts for the markedly greater effect of CO_2 heating in high northern latitudes. In the southern hemisphere the feedback is reduced by the regular seasonal break-up of the Antarctic pack-ice.

There have been many attempts to estimate the amount of heating needed to break up the permanent pack-ice of the Arctic Ocean, an enormous ice surface that has probably persisted for the past 700,000 years or more. If this ice were removed, or if it were confined to the winter and spring seasons, immense consequences would ensue. Quite apart from the obvious strategic and economic importance of the removal there is the fact that removal of the polar ice would imply a major reordering of world climate; the present-day pack-ice is a key controlling factor of the earth's radiative balance and general circulation of the atmosphere and oceans.

Manabe and Stouffer have estimated from their recent GCM experiments that a doubling of CO_2 would fail to remove the pack-ice in summer, but a quadrupling of CO_2 (probable in the twenty-second century) would do so. Thereafter in the average year there would be winter ice only, as at present in Hudson's Bay and on much of the Southern Ocean's Antarctic margin. If this estimate is correct, the next century is likely to a shrinking effect on the continental snow surfaces, and a trimming of the margins of the polar pack. There would be no clearance of the Arctic Ocean itself, though the ice would certainly become thinner. This view is opposed to that of Parkinson and Kellogg who found that a doubling of CO_2 concentration would indeed remove pack-ice in summer. Clearly there is a need to bring the assumptions and techniques of the modelers closer together.

Another point concerns the possible effect of a CO_2-induced warming on glacial ice, and hence on sea level. Contrary to public speculation a doubling of CO_2 is unlikely to produce drastic short-term effects on sea level. The predicted temperature rises will probably reduce the extent of alpine glaciers. Sea levels have recently been rising globally at rates generally estimated in the range of 10-15 cm per century, due partly to this process, and partly to expansion of the ocean water. But the main storage of ice on land is in the great continental glaciers of Antarctica and Greenland. The ice cover of Antarctica is very old, and losses of ice are almost entirely by means of ice discharge into the sea, a process that is not very temperature-sensitive at present temperature level. Greenland loses ice half by thermally induced ablation and stream flow and half by ice discharge, mostly through a single ice stream. A considerable increase of the marginal ablation would be likely to follow the CO_2 warming. On the other hand snowfall on the upper and middle slopes of both ice sheets would probably increase, thus altering the entire mass balance. During the next two centuries it seems unlikely that a large net transfer of water to the sea will occur, though the present slow rise will probably accelerate slowly. In the longer term the threat of significant sea level change from net glacial melting remains present. On the scale of many centuries it may well become important.

It is conceivable that catastrophic changes may affect the so-called West Antarctic ice sheet. Much of this ice is afloat or rests on a rock surface well below present sea level. Mercer, enlarging on a speculation of Hughes, suggested that the West Antarctic ice might disintegrate if a CO_2 warming becomes pronounced. Most glaciologists seem to question this opinion, but agree that should the ice disintegrate sea level would rise by about 5 m worldwide. The last stages of the disintegration might occur within a century. Catastrophic surges of the main East Antarctic ice sheet have also been hinted at. Hollin and Barry have pointed out that such surges are mechanical phenomena; only if CO_2 warming leads to increase of snowfall on the ice sheet, and if a warmer ocean trims back the margin of the ice sheet, is a climatically induced surge likely to occur. We conclude that catastrophic rises of sea level due to transfers of glacial ice to the ocean are unlikely, but not impossible. Within the next century it is improbable that major sea level changes will be a significant outcome of the warming.

Other Mechanisms

What other mechanisms may contribute to climatic change in the next two centuries? There is general agreement in the research community that the CO_2 effect will be the most significant influence, even though estimates of its magnitude differ widely. But several other mechanisms have been proposed, some of which may add to the CO_2 effect, and some reduce it. The list includes the effect of changes in the particle and aerosol concentration in the atmosphere, the effect of other infrared absorbing gases, and the possible effect of flip mechanisms in the atmospheric general circulation (the so-called almost-intransitivity effect).

There has been much speculation that variable concentrations of atmospheric particles may induce changes of surface temperatures. The primary source of effective particles, chiefly sulphates in the lower stratosphere, is considered to be explosive volcanic eruptions. Distinct coolings were, for example, observed in the wake of the eruptions of Tambora and Krakatoa. The 1982 eruption of El Chichón in Mexico has yet to show its effect, although the resulting stratospheric dust cloud has been widely observed.

There has been much debate as to the reality of the particle cooling. It has been argued theoretically that in some circumstances perhaps dominant the net effect of aerosol increase may be lower tropospheric and surface warming. Unfortunately, most of the parameters needed to settle the argument have yet to be routinely observed.

In two papers Bryson and Goodman calculated atmospheric turbidity since 1880 from measured attenuation of the direct solar beam. Their results showed a prolonged period of low turbidity, corresponding rather closely to a reduced number of large explosive eruptions. They showed a close relationship between northern hemispheric temperature and the turbidity; a clear atmosphere favored high surface temperature. The post-1938 cooling, in this view, resulted from the resumption of more active explosive eruptions. Their model even predicted the upturn since 1965. A rather different analysis by Schneider and Mass also found that the 1920 40 upturn was explicable in terms of dust loading, but that a reduced solar constant (due to sunspot activity) since 1950 may have contributed to the downturn.

Other recent attempts to include the effect of volcanic dust on global temperature have been undertaken to identify the modifying effect of such dust on the CO_2 signal calculated from models. Hansen *et al.* estimated the volcanic effect from Lamb's dust veil index, whereas Gilliland preferred to use the record of acidity from Greenland ice cores. When the volcanic effect was empirically weighted, and variations in solar luminosity were included, a close fit between the CO_2, solar and volcanic forcings and the observed variation of northern hemisphere temperature was obtained.

A factor that appears to reinforce the CO_2 heating is the presence in the atmosphere of numerous other gases that have infrared absorption bands in the `window' region the band between 7 and 14 μm wavelength in which the atmosphere is largely transparent to terrestrial radiation. Carbon dioxide and ozone both reduce the transparency in this band, and hence tend to raise surface temperatures. The effects are incorporated into the various models described above, as is the effect of H_2O. But there are other active gases present in currently negligible amounts that are being increased by human activity. Wang *et al.* showed that a doubling of the concentrations of nitrous oxide (N_2O), methane (CH_4) and ammonia (NH_3) could increase surface temperatures by about 1.1 °C, which is about half the expected CO_2 warming. The extensive use of agricultural fertilisers may indeed lead to increases in the abundance of some of these gases.

More recent synthesis has extended the list of such gases, and confirmed their importance by comparison with the effect of CO_2 acting alone. Hansen *et al.* calculated the equilibrium warming that increases in the chief absorbers may have contributed between 1970 and 1980, using a model that predicted a 3°C rise for doubled CO_2. They found that CO_2 may have raised global surface temperature by 0.14 °C in this decade, and the other absorbers (methane, nitrous oxide and two chlorofluoromethanes) by 0.10 °C. If this were representative, and if the other absorbers were to increase at significant rates in the next century, the CO_2 effect would clearly be enlarged. Unfortunately we only speculate as to such continued increases for the other absorbers.

The present and future states of the climatic system are only approximately known. There are grounds for supposing, as just shown, that rising CO_2 and trace gas concentrations in the atmosphere may tend to raise global surface temperatures in the next 80 years at an average rate of 0.06 °C a^{-1}, with maximum effect in high northern latitudes. It seems likely that this warming will affect precipitation and evaporation amounts. The uncertainties in the estimates have been deliberately stressed. Two questions now arise. Does such a change matter to humanity? And are surprise changes possible?

Seen in the perspective of the observed behavior of climate over the past century the above rate of change of temperature is impressive. Between 1880 and 1940 northern hemisphere temperature rose at about 0.01°C a^{-1}, during the last two decades of the period the rate being about 0.02°C a^{-1}. From 1938 until 1965 it fell at a comparable rate. The combined heating rate postulated for CO_2 and other infrared absorbers will eventually be much greater. These changes will all be perceived, in a highly variable climate, as changes in the incidence of extremes. A rapid change will get rapt attention, once it begins in earnest. On the other hand human societies have repeatedly absorbed even more rapid local or regional changes. Wittwer pointed out, for example, that Indiana in the United States had seen a rise of temperature of 2°C in the past century, and that from 1915 to 1945 the increase was at the rate of 0.1°C a^{-1}.

Changes of climate due to the gradual accumulation in the atmosphere of pollutants, or growth in the concentration of natural constituents like CO_2, are likely to lead to progressive but irregular changes of the sort just described. But more impulsive changes are common in nature, and these may present greater problems for human adaptation. These impulsive changes are often visible in local or regional precipitation records. Habitually they last only a few years, but in a few cases may extend to some decades. There are theoretical grounds for supposing that the general circulation of the atmosphere may flip from one quasi-stable mode of behavior to another with little change of external forcing although the argument does not justify the conclusion that local flips are of such origin.

Remarkable instances of such flips on the regional scale have recently affected many parts of the world's desert margins, where climate has been highly perturbed in recent decades. In the 1950s, for example, lake levels in East Africa were low, as they had been since the beginning of the century. Early in the 1960s 1961 in the case of Victoria Nyanza, and hence the headwaters of the White Nile there occurred an abrupt increase in rainfall over some of the plateaus. Some lake levels rose abruptly to levels not seen since the nineteenth century. There was no warning of this increase. High rainfalls persisted for several years in some areas, assisting in the spread of agriculture in the East African highlands. Desiccation returned after almost a decade of abundance, which misled many African statesmen into believing that a lasting change had occurred.

This East African anomaly was accompanied by a downward trend of rainfall along much of the southern margin of the Sahara. It, too, began in the early 1960s, and even earlier in some West African localities where, however, rainfall was well above average at the outset. Between 1968 and 1973 the desiccation was of such intensity that the fluctuation became known as the Sahelian drought. In fact it was part of a two to three-decade-long fluctuation of rainfall over all equatorial Africa, of which the high rainfall phase of the early 1960s was a part. Fluctuations like this are probably internal to the climatic system. Such unpredicted variations pose a threat of major disruptions to the human economy over large parts of the world.

References

Barry, R.G. & Chorley, R.J., (1992). *Atmosphere, Weather & Climate*, 6th edn. Routledge, London. 392pp.

Bradley, R.S. & Jones, P.D., (1992). *Climate Since A.D. 1500*. Routledge, London, 679pp.

Kemp, D.D., (1994). *Global environmental issues: a climatological approach*, 2nd ed. Routledge, London, 224pp.

Parry, M.L., (1990). *Climate Change & World Agriculture*. London: Earthscan Publications Ltd.

Pickering, K.T. & Owen, L.A., (1994). *An introduction to global environmental issues*. Routledge, London, 390pp.

Turner, R.K., (1995) *Sustainable Development and Climate Change*, Working Paper PA 95-01, Centre for Social & Economic Research on the Global Environment (CSERGE).

4

Adapting to Climate Change

The environment we see around us, in our towns and cities as well as in the countryside, is a result of current and past interactions between humankind and nature and it is constantly changing. However, due to climate change, as well as to demographic shifts and changing patterns of work and habitation, the pressures and demands on our environment as the century unfolds will be unprecedented. This will present both threats and opportunities to which we as a society will need to adapt.

At the same time, our natural environment is our greatest asset: the basis on which we can build a future in a rapidly changing climate. It provides the services that clean our air and water, and give us food, medicines, energy, and raw materials. Natural systems regenerate soils and pollinate crops, regulate the climate, cool urban areas, capture and store drinking water; and help to control floods.

Climate Change: The Risks of Inaction

Without action to adapt to a changing climate, some of the possible consequences for our natural environment include:

- Detrimental impacts on habitats and species, including changes to the composition of plant and animal communities and changes in the timings of seasonal events such as spring leaf growth, flowering, and breeding.

- Negative impacts on ecosystem services including: air quality, especially in urban areas which may be adversely affected by heatwaves; availability of water - reduced summer rainfall, particularly in Southern England, may affect the amount of water available for important biodiversity and habitats; and soil quality which will be influenced by changes in temperature, rainfall patterns and the increased frequency of extreme weather events.
- Difficulties for farmers and land managers including drought, leading to lack of water for crops and livestock, extreme hot weather, storms and heavy rainfall, causing increased soil erosion, water pollution, and storm damage. Subsequent changes to crop yields, different growing seasons, the introduction of new varieties of crops and livestock could all have major implications for our landscapes and the natural environment.
- Serious impacts on native woodlands and timber production driven by changes in water availability, particularly in the south and east of England towards the end of this century.

Increased frequency of flooding with consequences for ecosystems, agricultural production, people and businesses. Changing patterns of rainfall and sea-level rise will mean an increased risk of flooding and coastal erosion, and may also put pressure on cherished landscapes; Ocean acidification, caused by increased amounts of carbon dioxide, is a threat to many organisms in our seas.

A Strategic Approach

The natural environment underpins all aspects of our lives. It will be affected by climate change, and yet we will be increasingly reliant on it to help us manage the impacts that a changing climate will bring. It is a complex system that is affected by a vast range of people and organisations across all sectors and a range of pressures beyond those to do with the changing UK climate. This is a new kind of challenge and will require new approaches from all of us.

A Complex Challenge Requiring Integrated Solutions

As the climate changes we will need to develop mechanisms for managing changing demands and competing pressures on our natural resources. These will include opportunities to expand agricultural production, pressure to find space for new forms of energy generation, impacts on the quality and quantity of our water supply, and the demands of a growing population for housing, land for recreation and new infrastructure. In responding to these pressures, we must also consider how adaptation actions by some policy and decision-makers which may create benefits could also serve to complicate future impacts on the natural resources and services that other parts of society value.

The climate change risks and priorities that we are dealing with are highly interconnected and we need to ensure that we work in a way which seeks to reflect this. The resources and services that we receive from our environment cannot be adequately understood in isolation from one another. We will be able to find more efficient and effective solutions to the challenges posed by a changing climate if we consider individual parts of the system, such as water, biodiversity, agriculture and landscape as components of a complex and highly interconnected whole. We need to develop sustainable solutions that work effectively for people, the economy and the natural environment. There is a strong case for close and collaborative working with other experts, stakeholders and delivery partners to encourage a coherent response to the challenges ahead.

Working in Partnership to Deliver Adaptation

Climate change presents everyone – central government, local authorities, emergency services, other public sector organisations and the private sector – with an enormous challenge and also some important opportunities. Our water resources, health, coastal communities, cherished places and landscapes, even our food supply, face impacts from a changing climate that we need to prepare for. Government, local authorities, businesses and individuals all have a part to play in addressing this challenge. This means working in partnership at every level from local communities to the cross-Government Adapting to Climate Change Programme.

The natural environment – supporting the delivery of cross-Government policy outcomes The impacts of climate change on the natural environment have implications for a range of Government departments as well as Defra. Departmental Adaptation Plans (DAPs) are being published by major central government departments, and will be shared with their stakeholders and key delivery partners. In their Adaptation Plans, individual Government departments have highlighted the importance of a well-adapted natural environment to their agendas. The environment is recognised as playing a critical role in providing services that support the health and wellbeing of the population, and for vital underpinning of the whole economy. Across Government as a whole, the DAP process has served to highlight the natural environment as one of the most important areas for joint attention in the future.

> "Creating greener neighbourhoods and improving the assessment of populations and areas at risk can contribute to approaches that improve adaptation to the increased frequency of flooding and heat waves. This can prevent the onset of mental ill health related to poor quality housing and post-event stress."
>
> "Planting trees and vegetation and the creation of green spaces to enhance evaporation and shading are other options, as temperatures in and around green spaces can be several degrees lower than their surroundings." Department of Health
>
> "We are consulting on strengthened planning policy on a natural and healthy environment, bringing together policies on open and green spaces in rural and urban areas, and advising planning authorities to look more strategically at the functions green spaces can perform. The draft PPS requires local development frameworks to set out a strategic approach for the creation, protection and management of networks of green infrastructure, particularly in locations where it will assist in reducing the impacts of climate change by providing flood water storage areas, sustainable drainage systems, urban cooling and local access to shady outdoor space; and to avoid development being located in areas which result in the fragmentation or isolation of natural habitats." Communities and Local Government
>
> "DfT's Rail Division is working with the rail industry and Highways Agency to look at improving drainage systems including the use of Sustainable Urban Drainage Systems (SUDS) to be more resilient to increased rainfall. Management of the "soft estate (verges, embankments and un-trafficked paved areas) could be adapted to increase drainage capacity, reduce runoff, prevent flooding and landslides." Department for Transport

These initial plans are only a first step – they will be reviewed annually, more substantially in 2 years' time and again every 5 years, to ensure that further progress is being achieved in terms of government risk assessment, decision-making and capacity-building to deal with climate change.

The Environment Agency

Climate change is a corporate priority for the Environment Agency. All major delivery functions have undertaken a climate change risk assessment and considerable progress is being made. The Agency's Long Term Investment Strategy for Flood and Coastal Risk Management, which sets out the investment priorities for the next 25 years, is based on modelling using the UK Climate Change Projections 2009 (UKCP09). The Thames Estuary 2100 project has developed a tidal flood risk management plan to protect London from sea level rise, which is at the leading edge of adaptation policy. It is a strong example of how to implement the Green Book's Supplementary Guidance on Adaptation. The Agency has a Climate Change Training Programme being rolled out to all staff, with specific modules in development to increase knowledge and skills for particular roles, such as engineers, environment protection, flood risk, and water resources.

> "Climate change is already affecting the UK and the challenges we face as a result are only going to get tougher... Our new plans set out our approach to protect more people from flooding, clean up their local rivers, and help businesses reduce the resources they use. We all have our part to play to help protect the environment for future generations."

Forestry Commission

The Forestry Commission already has a comprehensive climate change research programme (largely undertaken by its research Agency, Forest Research). They have established a climate change communications initiative and commissioned the independent Read Report 'Combating Climate Change: a Role for UK Forests'. The Forestry Commission has also set up a new Centre for Forestry and Climate Change to study the effects of environmental and climate change on forestry, and how the sector can play its part in adaptation and mitigation of climate change. Key adaptation objectives for the Forestry Commission are: to increase the resilience of trees, woods and forests to climate change; to increase the role of trees and woodland in adapting the rural landscape to climate change; to enhance the role of street trees and urban woodland in minimising the impacts of climate change on our towns and cities; and, to use trees, woods and forests to help communicate and improve understanding of climate change issues and bring about behavioural change.

> "British woods and trees will not only play a key role in storing carbon and cutting the emissions causing global warming, but also help wildlife and people cope with impacts such as hotter summers and increased risk of flooding".

Joint Nature Conservation Committee

The Joint Nature Conservation Committee (JNCC) is the statutory adviser to Government on UK and international nature conservation, on behalf of the Council for Nature Conservation and the Countryside, the Countryside Council for Wales, Natural England and Scottish Natural Heritage. The JNCC's work contributes to maintaining and enriching biological diversity, conserving geological features and sustaining natural systems.

> "The natural environment is an integral part of the climate system, it is both affected by and effects climate globally and locally. To implement effective adaptation measures in the UK account of both needs to be taken. [...] It is important adaptation measures take an holistic, cross-sectoral approach bringing together different sectors. Taking an integrated approach to action may result in mutual benefits by reducing emissions, climate change impacts and biodiversity loss."

Natural England

Since its inception in October 2006 Natural England has placed sustainable climate change adaptation at the heart of its operational policy, decision making and advice. In March 2009, Natural England published assessment of landscape scale climate change adaptation options for the Norfolk Broads, Cumbria High Fells, Shropshire Hills and Dorset Downs. The project has now been extended across England into a number of other character areas and regional scale studies. Natural England manages approximately £0.5 billion a year of agri-environment funding to farmers and other private land managers and have integrated climate change adaptation and mitigation objectives into Environmental Stewardship schemes. Natural England is developing a coordinated programme to train staff, assess threats and opportunities from climate change across all its areas of work, and to identify opportunities to contribute to wider adaptation and mitigation goals. Natural England has a climate change evidence programme, researching and developing adaptation and mitigation, working with Defra and other partners.

Understanding the Issues

Climate change will add to the pressures we already place on natural resources whilst increasing our reliance on the services we receive from them. This will present us with difficult decisions to make about how to adapt and how to cope with unavoidable change. It is important that we are honest about these impacts, frank about the uncertainties that we face, and therefore flexible in our responses. Here we set out our understanding of the challenges, and clarify Defra's approach to addressing change across the range of our priorities.

1. Change is Unavoidable

Our natural environment is changing. However much we cut our greenhouse gas emissions now and in the future, we can expect at least 30-40 years of temperature rise due to past human activity. The speed and scale of these changes will be unprecedented, presenting both threats and opportunities. Direct impacts will be seen most clearly in our vital natural systems.

We can expect warmer, wetter winters; hotter, drier summers, warmer more acidic seas; and an increase in the frequency of extreme weather events, resulting in changes to species distribution and habitats. Valued landscapes will change and some features may be lost or significantly altered. Both mitigation activities and adaptive actions in other sectors may increase pressures on the natural environment. In some cases these associated impacts are likely to be seen in advance of the more direct impacts of climate change.

The natural environment will make its own adjustments in response to changes in climate. However, sustaining and optimising the benefits we receive from ecosystems (from the aesthetic qualities of our countryside to the provision of healthy food and clean water) will require some human intervention. Difficult decisions about priorities for future management of land and other natural resources will need to be made. These will need to be informed by the best available science but must also respond to the views of the communities of place and interest that value different aspects of our natural environment.

2. Adaptive Capability of the Economy and Society

It is increasingly well recognised that our quality of life – from the 'livability' of our communities to the stability of our economy – is dependent upon maintaining healthy natural systems. In planning our response to the social and economic challenges posed by a changing climate we need to understand and capitalise on the crucial role played by the natural environment.

Well planned 'green infrastructure' in the form of green spaces, street trees, modern, sustainable drainage systems, and open waterways can help us reduce and manage the impacts of heat waves and surface water flooding in urban environments as well as providing a range of additional social, health and economic benefits. On key issues like access to clean water during hotter, drier summers or managing flood risk in the face of increased numbers of extreme weather events; we can't respond effectively to projected climate pressures without effective environmental management, which balances the multiple demands we place on ecosystems. An understanding of the value of these ecosystem services will be central to our capacity to manage the risks associated with climate change.

We have passed the stage where we can behave as though there were a choice between the environment and economic growth or improving social outcomes. Living sustainably means pursuing all three goals together, recognising that safeguarding our natural environment is now absolutely vital for the future of our economy and society.

3. Non-climate Pressures will Continue to Drive Change

In the short and medium term existing, non-climate pressures on the natural environment, such as population change, housing growth, marine development and the provision of infrastructure to support growth, will continue to be significant and will need to be addressed alongside the impacts of climate change.

Over time, these pressures are likely to be exacerbated by climate change, for example population and housing growth may create additional demand for water at a time when summer water availability is decreasing – particularly in the most densely populated parts of the country. Similarly, there will be ongoing competition between different potential land uses – such as housing, food, renewables, recreation, managing flood risk and biodiversity.We will need to develop decision-making tools which support the active management of these tensions. The facts about climate change may mean that we need to re-think the costs and benefits of certain courses of action and reconsider priorities in the light of new information about risks.

We will always need to strike a balance between space for nature and the other multiple demands that society places upon land. What is increasingly clear however, is that in future this balance will need to more accurately reflect the true value of the services provided by a healthy natural environment.

4. Environmental Quality Remains a Viable Goal

Many of our existing definitions of success were developed on the basis of limiting and, where possible, reversing harm done to the environment over the past century. However, we now know that over the long term our natural environment will be quite different to anything we have seen to date. If we accept that maintaining the status quo or restoring the environment to an historic state are no longer realistic options we will need to rethink the way we approach these goals.

The challenge will be to recognise the changing context brought about by climate change whilst maintaining continuity of purpose in seeking opportunities to improve environmental quality. This means using our improved evidence base and building on existing commitments to safeguard and enhance our future environment.

We cannot control the direct impact of changes in climate on landscapes, habitats and species, but we can make active decisions about how we utilise our environment and where to intervene, recognising the key role played by healthy natural systems in underpinning society, supporting the economy and enabling adaptation. We will need to develop a more flexible approach to policy and delivery that allows us to respond to issues as they unfold and set this within a clear overarching vision of functioning, sustainable ecosystems.

Difficult tradeoffs will continue to be made at all levels but these can be managed in a context that continues to actively seek net improvements to the quality of our environment. We need to accept the reality of change but we must also recognise that crucial decisions about the quality of the future natural environment remain in our hands and relate directly to our future quality of life.

Future Natural Environment

Our natural environment has changed considerably over the last two centuries because of industrialisation, urbanisation, changing farming practices, improved transport infrastructure and now climate change. Climate change is likely to accelerate these changes in the coming decades with impacts for the environment, people and the economy. However, it is also important to recognise that within this changing context our environment will retain many valued qualities and may also develop new features that we wish to protect and enhance.

A key component of adaptation will be to aid people's understanding of the drivers of change and to enable people to think about the opportunities as well as the threats. The land management and conservation decisions we take in response to climate change will need to reflect a range of factors - important amongst these are the views of the people who value the natural environment.

Climate Change Impacts on Landscape Character Areas

Climate change will have a range of direct and indirect consequences for different aspects of the natural environment. These impacts can be best considered and addressed by looking at the likely vulnerability of the natural environment, and developing adaptation strategies, at a 'landscape scale', i.e. across a sufficiently large geographic area to be able to consider large scale natural ecosystem processes, the way people use and value an area, and the interactions between them. The aim of such an approach is to sustain the full range of benefits we receive from the natural environment.

Natural England has adopted this approach by studying four of England's 159 National Character Areas in detail, assessing the potential impacts of climate change and setting out some possible management responses.

Cumbria High Fells: a mountainous landscape likely to be affected by an increase in temperature and an increase in winter rainfall. Impacts are likely to include a move of species towards north facing slopes and higher ground, with those on the highest tops being at risk of extinction; peat bogs drying out and consequential increased carbon losses to the atmosphere; reduction in lake levels and increasing lake & river water temperatures. Footpaths being eroded by a combination of increased drought, recreational pressure and intense rainfall events. Adaptive actions will include encouraging native woodland regeneration in higher areas to help species migration, the restoration of peat land, a more holistic approach to water management in the uplands, and continuing projects which improve the resilience of footpaths, such as 'Fix the Fells'.

The Norfolk Broads: a low lying wetland landscape which is likely to be affected by coastal erosion and sea level rise. Impacts on the Broads are likely to be more significant than for some of the other areas featured in this work as salt water increasingly intrudes into the freshwater lakes, causing habitat change and species loss, and due to the increased likelihood of both flood and drought. Adaptive actions might include revision of river dredging regimes, planting wet woodland in flood plains and restricting recreation at times of poor water quality. Working closely with regional and local partners the Broads Authority is currently revising the Broads Plan to take account of climate change and the need for adaptation. This will involve a public consultation and look at social, economic and environmental impacts.

Shropshire Hills: a farmed landscape that is likely to be affected by higher temperatures and less rain in summer (but more in winter). Impacts here include a loss of mature trees in the landscape as these succumb to extended droughts and more severe storms; changes in the viability of some crop varieties and livestock breeds; damage to historic buildings, caused by an increase in soil erosion during peak rainfall events. Adaptive actions include extending existing areas of semi-natural habitat and creating new areas; promoting a variety of tree species to eventually replace existing mature trees.

Dorset Downs and Cranborne Chase: a chalk landscape which is likely to face drought. If we do not reduce greenhouse gas emissions, by 2080 the climate of the area may resemble parts of Portugal. The natural communities that currently characterise the landscape will change. For example, the shallow rooting beech, which is common in Dorset is likely to decline, but small leaved lime needs warmth to set seed and will probably be able to increase. Veteran trees of all species are more likely to be felled by storms. However, in woods the impact of these storms can also be positive, creating glades that species adapted to sunlight can occupy. Adaptive actions include re-establishing chalk grassland or native woodland adjoining water courses to reduce downstream flooding; and creating naturally functioning floodplains to allow greater water storage and the evolution of new wetland habitats. Following the completion of the study, Natural England has produced a video in which a resilient natural landscape is described, in pictures and words, by a range of partners.

Adaptation Action Needs to be Sustainable

This means that our responses should not add to climate change, or limit the ability of other parts of the natural environment, society or business to carry out adaptation elsewhere. Our responses must avoid any detrimental impacts on other parts of society, the economy or the natural environment.

Impacts of Adaptation in the Farming Sector

A good example of these crucial inter-linkages between society, business and the natural environment is the farming sector, which is amongst the first to experience the direct effects of a changing climate. A recent survey indicates that over half of farmers believe that climate change is having an effect on their farm now and nearly a third are already taking action to adapt. Agricultural land represents around 75% of the UK and provides a wide range of environmental, social and economic benefits to society with an influence that goes well beyond its primary purpose of producing food and non-food crops. These benefits include employment and other contributions to thriving rural communities; essential environmental processes such as water cycling and purification, maintaining air quality, reducing erosion and flooding, carbon storage and climate regulation; habitats for wild flora and fauna; and land for recreation.

"The productivity of agriculture must be enhanced while simultaneously reducing its environmental burden. This requires new investments in technologies, knowledge and skills to improve the future sustainability of agricultural land use. This will require diverse collaborations amongst many stakeholders, public and private, with interests in the future of land and the services it provides".

Land Use Futures

Some impacts will have a direct effect on both farming and the natural environment, e.g. flooding caused by increased storms and heavy rainfall - but there are likely to be secondary impacts on the natural environment as a result of measures taken by farmers to adapt to a changing climate, such as the introduction of new crops and livestock varieties, which may have either positive or negative effects on ecosystems and landscapes.

These risks may be compounded by other changes to agricultural practices, for example pressures to increase production to feed a growing global population or to grow more fuel crops. Therefore adaptation by agriculture will be managed most effectively by taking a whole-system approach and addressing production, conservation of the natural environment, and social and economic benefits in an integrated way. In a small, relatively crowded country where agriculture occupies three quarters of the land, an understanding of the wider picture will be essential if we are to avoid solving one set of problems at the cost of making others worse.

We therefore need to all work together – central Government and delivery bodies, local Government, farmers, land managers and NGOs – to manage adaptation.

Climate change impact	***Possible response by agriculture***	***Risk to natural environment of response***
Warmer conditions and longer growing seasons	Increased yields and the opportunity to grow new crops or existing crops further north. Introducing new livestock varieties.	Changes in wild plant and animal communities reliant on previous crops/livestock, impacts on soil and water quality from new crop types, potential new pests and diseases, and changes to landscape.
	Harvesting/planting earlier and producing two crops.	More intensive production and greater use of energy, fertiliser and pesticides, would have a detrimental impact on biodiversity, habitats and water quality.
Drought	Over abstraction of water for crops and livestock.	Impacts on water quality and availability and associated impacts on habitats and biodiversity.
	Planting more drought-tolerant species.	Changes in wild plant and animal communities reliant on previous species. Less biodiversity and habitat types if significant move to growing small number of tolerant species.
Extreme hot weather	Over abstraction of water to cope with fires.	Reduced water availability and associated impacts on habitats and biodiversity.
	Building more animal shelter, refrigerated stores and fitting vehicles with air conditioning.	Unless sustainable energy sources are used, greater energy consumption will increase emissions with repercussions for the natural environment.
Storms and heavy rainfall	Abandonment of land at risk of flooding.	Impacts on biodiversity and habitats from abandoning land.
	Using farmland for flood storage.	Impacts on biodiversity and habitats.
	More animal housing.	More building increases flooding risk and reduces space for wildlife.

Green Infrastructure

At the same time, appropriate land management can provide some of the most sustainable and cost effective ways of helping to protect society from the effects of a changing climate, as well as benefiting agricultural production. For example:

- Trees and other vegetation which sequester carbon, thus helping to mitigate climate change, also have a number of adaptation benefits. Trees can provide shade and shelter for crops, buildings and livestock; reduce runoff and soil erosion and slow the movement of floodwater; help to maintain water and habitat quality through shading and by filtering pollutants; support biodiversity including crop pollinators; as well as providing an alternative or additional crop (for renewable fuel or timber) to increase farm income and hedge bets against climate risks. Increasing woodland cover can also have recreation and cultural benefits.
- Sustainable drainage systems (SUDs) - which include porous surfaces, infiltration trenches, filter drains, ponds and wetlands, and grass buffers- slow the flow of water and increase its infiltration into the soil. This can help to both capture and store water at wet times of year to reduce the risk of shortages during drier times and reduce the risk of flood damage to areas downstream. It also benefits biodiversity.
- Peat restoration, especially in upland blanket mires, helps maintain huge stores of carbon, and can help to improve water quality and supply, and when in good condition may reduce flood risk, in areas downstream. It also has significant benefits for the conservation of biodiversity, and helps to maintain a valued landscape that contributes to a well- developed and important tourist industry.

Environment Supporting Quality of Life in Urban Areas

Natural England has emphasised that green infrastructure doesn't stop at the city limits – at its best it links across a network of green spaces and corridors, rivers and waterways through urban areas and the urban fringe and out into the wider countryside. It helps break down the barriers between people and the natural environment.

Well-designed and managed green infrastructure can deliver a wide range of environmental and quality of life benefits in towns and cities. Trees, parks and other green spaces, green roofs and walls and water – rivers and waterways, lakes and ponds – can all help to cool urban areas, counter the urban heat island effect, and help to address urban flood risk. Trees have an important part to play, for example in sustainable drainage systems (SUDs), while targeted floodplain woodland creation upstream of our towns and cities can help to delay and reduce peak flood flows. Green infrastructure can assist in providing cleaner air by filtering dust and harmful chemicals; roadside trees can trap up to 90% of traffic-related airborne dust particles. It can also increase sustainability by lowering energy costs (through reducing our reliance on air conditioning), and reducing the costs of tackling health problems during heatwaves.

Finding more ways to work with nature and provide essential green infrastructure will help to buffer the country as a whole from the effects of climate change. In many cases these are the most cost effective and efficient adaptation options currently available to us as well as delivering a wide range of additional social and economic benefits. The multiple benefits from good land management, and the role of farmers, land managers and place shapers in delivering sustainable adaptation, will become increasingly important as the climate continues to change.

Bedfordshire and Luton Green Infrastructure Consortium

The population of Bedfordshire and Luton is set to grow significantly over the coming years. The Bedfordshire and Luton Green Infrastructure Consortium is a partnership which includes local authorities and voluntary organisations as well as the Forestry Commission, the Environment Agency and Natural England. It has identified a need for green infrastructure planning and delivery in order to protect and sustain the environment for existing as well as future generations to enjoy.

The pressures on the area will include a need for an increased supply of affordable housing, as well as improved transport infrastructure and issues around urban deprivation and regeneration. The partnership recognises the potential pressures of climate change on the environment; as well as the role it can play in helping to adapt to potential impacts, through flood storage or by using woodlands and forestry as well as parkland to counter the "heat island effect". The partnership have identified 'green corridors' which will create a green infrastructure network across the region. The network is made up of urban and country parks, amenity open spaces, natural and semi-natural habitat for wildlife, nature reserves, Sites of Special Scientific Interest, designated landscapes, heritage sites and public rights of way. It also incorporates better connected cycleways and footpaths to enable healthy exercise and environmentally sound travel. Projects that are currently underway to deliver this green infrastructure network include Bedford's Green Gateway, the Chalk Arc, the Bedford Green Wheel, Bedford River Valley Park, Bedford River Corridor and Bedford and Milton Keynes Waterway Park.

Policy and Practical Challenges for the Future

Identifying and managing risks of mal-adaptation There is wide-spread recognition of the potential for adaptation actions that benefit one sector or part of society to have unintended impacts for other sectors or parts of society. It is easy to agree that adaptation actions must be sustainable but much harder to forecast the potentially negative consequences of changes in behaviour. In particular, it will be challenging to assess the cumulative impacts of a wide range of new trends and approaches. We are in the early phases of what is likely to be a period of experimentation in the land management, farming and conservation sectors. Across all of these sectors, there will be a need to build capacity for capturing and evaluating the impact of climate-driven changes in practice. An ability to do this better will enable promotion of the most effective approaches as well as alleviation of negative impacts.

Developing new skills sets to deliver innovative approaches A changing climate will create a demand for a range of new skills and competencies in delivering adaptation strategies. For example, delivering a network of green infrastructure capable of supporting climate change adaptation will require new skills in planning, design and management as well as more community involvement and empowerment. 'World Class Places' – the Government's strategy for improving the quality of place (May 2009) identified green infrastructure as one of the four elements of quality places and identified the need for a step-change in the provision of green infrastructure to help meet the challenges of climate change. Likewise in Grey to Green, the Commission for Architecture and the Built Environment (CABE) highlighted a gap in the skills needed to green our cities to help adapt to climate change and improve people's quality of life.

Evidence from local authorities suggests they may lack the key skills to deliver effective green infrastructure. Other evidence highlights labour shortages of over 90% in landscape architecture and urban design sectors.

Action should be Flexible

Although there is still uncertainty over the future climate, we need to consider options now and make decisions that maximise future flexibility - in many cases it is failure to take decisions that locks us into inflexible pathways.

Conserving Biodiversity in A Changing Climate

Biodiversity is a strong example of an area where we need to be prepared to act in the short-term in order to avoid limiting our options in the future. We may not know the precise effects, the exact costs or the definite timing of climate impacts, but we already know enough to start taking action. Climate change means that we need to adopt an increasingly dynamic approach to managing ecosystems and conserving our biodiversity. In 2008 the England Biodiversity Group published a set of adaptation principles, designed to help guide the approach to conservation of biodiversity:

- Take practical action now;
- Maintain and enhance ecological resilience;
- Accommodate change;
- Integrate action across all partners and sectors;
- Develop knowledge and plan strategically.

There is a need to ensure that biodiversity, habitats and ecosystems are in the best possible condition, so that they are able to continue to function effectively in the face of new or increased pressures.

We must continue our efforts to conserve and manage protected sites to bring them into favourable condition, as well as taking a landscape-scale approach, building in habitat enhancements to reduce other pressures on biodiversity and developing habitat networks to aid the dispersal of species. The challenge is to take the right action at the right time to minimise losses, facilitate change and take advantage of new opportunities.

One of the key challenges we have to face is designing conservation targets and objectives for the future that are demanding, but also achievable and sufficiently flexible to cope with the unpredictable nature of both climate change and society's response to it. The EU has a target to halt the decline in biodiversity by 2010 and the European Council re-stated its commitment to this goal in 2008. On the world stage, a global target exists to reduce the rate of biodiversity loss by 2010. New targets at both the European and global level are expected to be negotiated this year, showing an ongoing commitment to biodiversity conservation and enhancement.

It is possible that as the climate changes, sites that were designated for one set of species may not retain them at the previous levels of abundance, regardless of the level of conservation effort, even though those protected sites will remain a vital resource for many other valuable species and may become part of the distribution ranges for other protected species. It is possible that some iconic species or those that currently characterise a particular habitat may change their distribution or frequency too. In future, therefore, the emphasis is likely to be on improving and buffering the capacity of existing sites to cater for a wider range of species. Reducing fragmentation of habitats will also allow some species to move around the landscape more easily and find patches of suitable conditions.

There is a case for reconsidering traditional approaches in the face of substantial future challenges. Re-examining how we frame conservation objectives for individual sites could be needed, as well as different perspectives on governance. In future, our targets may need to be less informed by a historical perspective and perhaps more focussed on managing habitat condition so that sites are better able to withstand current and future stresses, enable the movement of species where necessary and continue to provide vital ecosystem services.

Porlock Ridge – an Example of Adaptive Management

An important aspect of adaptation to climate change will be the need to accept change and reconsider current practices. At Porlock Ridge, an SSSI on the Severn Estuary, adaptive management means that one high quality habitat is being naturally replaced with another. This shows how change can be accommodated successfully so that while one set of valued assets may be lost a new set is created.

A storm in 1990 breached the shingle ridge separating the marsh, which was designated as an SSSI, from the sea and the marsh was inundated with seawater. The damage was repaired but – recognising the likelihood that ridge would continue to be breached in future – work was commissioned to consider a sustainable long-term solution. Since a further storm in 1996 Natural England have initiated a "managed retreat" on the site which will slowly allow the land to return to the way it was 200 years ago, with a low, wide ridge of shingle and a lagoon behind. Land use has changed – the area is now grazed very much less, and the coastal path has been formally diverted. Birds and flora are being monitored, as is the changing shape of the ridge.

The previous habitat of Porlock Marsh has now largely been destroyed by inundation, but a new saltmarsh has been created which is one of largest on Severn Estuary. This is first time an SSSI has been renotified after natural coastal processes have operated and changed an area.

Flexibility and Diversity in Woods and Forests

Another good example of an area where we need to take decisions in the short-term with a view to maintaining flexibility and retaining benefits in the long-term is forestry. Because trees have a long lifespan any adaptive measures need to be appropriate to both current and future climatic conditions. By the end of the century, some native tree species are likely to lose 'climate space', particularly in southern England. The range and ability of the majority of priority woodland species (both flora and fauna) to persist will change as a result of climate change; some will decline further, others will benefit.

Many uncertainties remain about both the future climate and the response of trees and woodland ecosystems to climate change. Expanding the range of tree species to increase resilience – at stand or landscape scale – when restocking existing woodlands or planting new ones, represents the most appropriate adaptation response at this time. This type of 'no-regrets' approach can also provide additional insurance against the uncertain future impacts of pests and diseases.

The Forestry Commission is preparing a Climate Change Action Plan for the Public Forest Estate in England. The Plan benchmarks existing management practice and will initiate actions to enhance species diversity and genetic diversity through appropriate provenance mixtures. Changes in woodland management systems will also be considered, including those that promote diversity in age and structure and promote natural regeneration. The Plan will also ensure sufficient diversity within timber species at landscape and, in some cases, stand level. It will also require that less common and some relatively untried species are planted – in part a response to the reduced range of 'conventional' species as a result of recent pest and disease outbreaks. This will require that sub-optimally performing species (under current climatic conditions) are planted to provide the required diversity and, as a consequence, resilience to changing climatic conditions or the threat posed by pests and diseases.

Balancing a flexible approach with continuity of focus A key challenge for the future, for all parties with an interest in conserving the natural environment, will be developing flexible policies and setting targets that can adapt to reflect changing contexts; while continuing to recognise the value of biodiversity and ecosystems and their contribution to wellbeing and prosperity. Wildlife is under threat for a number of reasons – such as pollution, habitat loss and development and climate change can exacerbate all of these pressures.

It is important, however, that we are able to distinguish between present threats, pressures that are likely to develop in the short or medium term and those which may only develop under certain scenarios in the longer term. While we will have to accept that some changes are unavoidable, we will also need to develop capacity to distinguish between: avoidable climate-driven changes, avoidable changes driven by other pressures and those changes that we cannot realistically avoid. In all of this we need to continue our work to protect biodiversity so that we can continue to benefit from the ecosystem services our natural environment delivers.

Marine and Coastal Access Act

The Marine and Coastal Access Act 2009, the first of its kind anywhere in the world, introduces a single piece of legislation for the protection of the marine environment. Through a new marine planning system we will be able to manage our environmental, social and economic needs together, and an ecologically coherent network of Marine Protected Areas (MPAs) will be created by 2012.

Marine Conservation Zones are a new type of MPA. They will be designated to conserve flora, fauna, habitats whether rare or threatened or representative of UK biodiversity, as well as geomorphological features. Levels of protections will be site specific and dependent upon a site's conservation objectives. These are expected to range from sites which have little or no restrictions to sites where all damaging activities are excluded.

Existing Marine Nature Reserves are converted into MCZs under the Act, Lundy became and MCZ on 12 January 2010.

Our seas are already showing the effects of climate change and with increasing use of the sea by many competing interests, we must make sure that the marine environment can cope with changing conditions. We have a duty to look after our seas for future generations. It is envisaged that our network of MPAs will contribute to halting the decline in biodiversity by conserving representative UK habitats and species, and areas where there are rare and threatened species and habitats. Marine Conservation Zones are a flexible mechanism. They can be decommissioned, modified, or moved to ensure they are still protecting the feature they were designated to protect, or to ensure they continue to fulfil their contribution to the network. This is particularly relevant in relation to climate change. The UK network of MPA will contribute to protecting UK biodiversity, providing some respite from other pressures in the face of climate change.

Recognising Opportunities as well as Threats

Climate change is unquestionably a threat to the natural environment. Consequently, much of the focus is, quite rightly, on managing the risks and enhancing resilience. It is important however that we recognise the opportunities where they do occur. Climate change could be beneficial for some species in Britain allowing them to spread northwards and increase their range. A well-known example of this is the spread of Dartford Warbler, which used to be confined to small parts of Dorset due to cold winters, but which are now found as far afield as Suffolk and Staffordshire.

Given that some native species may become threatened in southern parts of their present-day distributions, taking a flexible approach may include welcoming the establishment in the UK of new species. For example the Little Egret has become a welcome sight in the south since its arrival in the UK in the late 1990s. We will need to make such judgements on a case by case basis, using sound science. It will be important to monitor the situation and control measures may become necessary where a non-native species does become invasive and a threat to conservation objectives.

Defra's Evidence Investment Strategy

The major global challenges of climate change, protecting ecosystem services and ensuring a 'sustainable food supply call for quality evidence and innovation to contribute to good policy making and sound government. Defra recognises this and invests heavily in evidence: approximately £240m to evidence-related activities, including research, surveillance and scientific specialists in 2009/10.

> "It is absolutely essential that all the policies that we formulate are based on sound evidence. We now understand more about the challenges facing the earth's climate, ecosystem services and the supply of sustainable and healthy food. There has never been a time when there was a greater need for good quality evidence to contribute to policy making and sound decisions."

There have always been compelling arguments for protecting individual ecosystems on the basis of their aesthetic and cultural values. However, Defra's new Evidence Investment Strategy (EIS) published in January 2010 emphasises that it is becoming increasingly clear how our natural environment offers more tangible benefits to the economy, lifestyles and even survival.

The strategy highlights protecting ecosystem services as one of the three big interlinked evidence challenges for the Department and recommends that more evidence investment needs to be directed to these big enduring challenges. Not only will habitats rand species be affected directly by climate change and sea level rise, they will also be affected by policy and behavioural shifts in sectors such as agriculture, water, transport and energy. While the general principles and direction of climate change impacts on biodiversity in the UK are understood, the details and timing of impacts on individual species, habitats and sites are very uncertain. Key areas identified for further evidence gathering include the implications of climate change for species (including non-native species) and ecosystems, and on wildlife disease. This is a long-term agenda, requiring a more sophisticated understanding of the value of ecosystem services and the relationship between economic and environmental performance.

Dealing with Uncertainty

A challenge – for all organisations – is to plan for the uncertainties that still remain over the impacts of climate change and to keep options flexible if the climate or other risks change faster, or more slowly than we believe they will at the moment. UKCP09 shows a range of climate changes up until the end of the century with associated probabilities to give a measure of the strength of evidence for each outcome, based on current science from the Met Office Hadley Centre, and including information from observations and other climate models.

UKCP09 represents a comprehensive survey of the spread of plausible outcomes consistent with current understanding, and illustrates clearly that it would be wrong to plan a strategy to deal with just one single projection (such as the central estimate for the medium emissions scenario). In addition, the projections are likely to be updated in the future as understanding and modelling technology improves. Therefore adaptation plans will need to be robust to known uncertainties captured in UKCP09, and also sufficiently flexible to allow modification in the light of new projections as they emerge.

Getting to grips with both the temporal and spatial dimensions of climate change It is clear that climate impacts will be felt differently in different parts of the country and that the impacts will change over time. A common challenge for all organisations seeking to assess and manage climate change risks based on sound evidence will be to move from a broad appreciation of the general impacts (warmer, wetter winters, hotter drier summers etc.) to a more specific understanding of the probable impacts on key outcomes and objectives in a given area over a given timescale. This will be important in ensuring that adaptation responses are both timely and proportionate.

Building capacity and capability to plan ahead Whilst continuing to build the evidence base and improve the modelling are crucial elements in building our capacity to adapt, it should be recognised that these are only part of a wider story. Understanding how best to work with probability and projected scenarios, how to establish priorities based on an effective analysis of risks and how to plan for the long term all pose challenges for individuals and organisations in all sectors. All of us will need to improve our capacity and capability to work in ways which may be new to many of us. This will be an iterative process that will require a culture of ongoing learning and a willingness to share and evaluate new approaches.

Action Research

Whilst we have an increasingly sophisticated evidence base addressing the potential impacts of climate change both spatially and over time, much less is known about the factors influencing the capacity of organisations and individuals to respond to this information. Defra is currently undertaking research to improve our understanding of both the practical processes and the strategic choices and challenges involved in applying the UK Climate Projections to land management for environmental purposes.

The project aims to enhance our understanding of the full range of factors influencing land management organisations seeking to incorporate climate risks into their planning processes. The project focuses on a package of action-learning delivered to four case study organisations.

Clinton Devon Estates – focussing on Clinton Farms to provide an insight into climate change adaptation with an environmental focus on agricultural land. Environmental considerations are a key priority, including heathland management, river management and forestry. Concerns for the future include impacts on forests and woodland as a result of climate change and the potential for an increase in exotic diseases and pests.

Lake District National Park – focussing on the Windermere Catchment Restoration Programme; a partnership programme involving a variety of other organisations including the Environment Agency. Their key concern is to improve water quality and preserve the natural ecology of the catchment and lakes, which in turn will help to support a healthy, local economy. Climate change is a key concern in improving land management plans to take into consideration potential future risks including increased levels of flooding and erosion.

Lea Valley Regional Park Authority seeks to meet the leisure, recreation and nature conservation needs of London, Hertfordshire and Essex. Focusing on Walthamstow Marshes on the margins of east London, an SSSI and one of London's last natural floodplain grasslands. Climate related concerns include issues around too much water causing flooding; too little water from drought or over abstraction; water quality which affects users for consumption and leisure (e.g. canoeing) and an increase in visitor numbers (as a result of warmer summers) leading to environmental pressures.

Royal Society for the Protection of Birds – the RSPB promotes conservation and the protection of birds and the wider environment as well as managing 200 nature reserves. Focusing on Dearne Valley Reserve near Barnsley in South Yorkshire. Climate related concerns include cost and damage to the environment as a result of invasive species and changing migration patterns. Changing rainfall patterns may be exaggerated within the Dearne Valley as it receives floodwater for the surrounding urban areas leading to additional concerns around flood defence and increasing spring droughts.

Managing Flood and Coastal Erosion Risks

Flooding and coastal erosion already cause significant damage and disruption in England. Changes in sea level and weather patterns are expected to increase the frequency and intensity of all types of flooding and coastal erosion with impacts for people and the economy. These will also pose new challenges for the natural environment. Increased flooding could result in more soil, debris and pollutants being washed into watercourses, with consequences for biodiversity. Sea level rise against fixed defences could lead to the loss of ecologically important coastal habitats such as intertidal flats and marshes and threaten the sustainability of supra-tidal habitats such as sand dunes and shingle banks. This will also increase the risks as these habitats lose their capacity to break the destructive force of wave energy reaching the coast. Saltwater flooding may adversely affect vulnerable freshwater habitats at the coast. Action to meet these challenges needs to be planned well in advance.

Different approaches to flood risk management will be needed in the future, including those that work with nature. Hard defences to reduce the risk of flooding and erosion (such as concrete barriers) cannot easily adapt to rising sea levels and may make erosion worse. Also, they could impact on biodiversity by altering the physical properties (hydromorphology) and diversity of river corridors, wetlands and coastal areas. Faster coastal erosion could lead to destruction of sewers or exposure of old rubbish dumps, with consequences for human health and the environment. And loss of development land to inundation or erosion could place further pressures on the environment, as we seek to move or replace homes, roads and other infrastructure.

There are also opportunities to manage the increasing risks in ways which both enhance the environment and mitigate the impacts on people. For example:

- Creating "blue corridors" to make space for water within urban areas could have wildlife and amenity benefits - Defra is researching the viability and benefits of this approach.
- The realignment of defences to more sustainable locations could create ecologically valuable intertidal areas and fluvial habitats which also

help dissipate wave energy and reduce the risks to people.

- Restoring wildlife rich wetlands and natural flood retention areas could reduce flood risk downstream.
- Restoring habitats such as woodland in catchments and improving land and soil management could reduce run-off and local flood damages as well as sustaining the capacity of land to support agriculture in the long term.
- Increasing the use of Sustainable Drainage Systems (SUDS) can improve water quality and public amenity, improve the attractiveness of urban developments through the incorporation of ponds and reedbeds, and help to tackle flooding from surface water run-off by reducing the volume and flow of water to sewers. SUDS also offer biodiversity benefits by creating habitats for biodiversity.

In future a mix of both 'hard' and 'soft' approaches to managing flood risk and coastal erosion are likely to be required to ensure we meet the needs of people, businesses and the wider environment. This will need to take account of both the nature of the risks involved and the full range of costs and benefits associated with alternative options.

Water—Managing Multiple Demands

Climate change will affect the spatial and temporal distribution of rainfall in this country. Summer rainfall will reduce, particularly in Southern England, and less water may be available for abstraction, e.g. from rivers for irrigation. Recent work by the Environment Agency suggests that the net effects of changing rainfall patterns will reduce river flows by 15% by the 2050s. Climate projections also suggest that winter rainfall could increase in some regions by as much as 30% by the 2080s, while rainfall intensity could increase both in winter and summer. In England, our public water supply systems take advantage of rainfall from October to March to replenish reservoirs and aquifers, with relatively little dependence on summer rainfall. However, agriculture tends to rely on summer rainfall for crop growth.

Because water is essential for life, changes in rainfall distribution have wide-ranging implications for the natural environment:

- Water quality and aquatic flora and fauna could be affected as reduced river flows will make it harder to dilute or disperse pollutants.

- A lack of water may affect fragile habitats such as chalk downs leading to changes in species and their distribution.
- Farmers and businesses may suffer through a lack of water for food production and processing.
- An increased frequency and severity of summer drought is likely to represent the greatest threat to woodlands from climate change.
- Low flows could affect tourism on our waterways by preventing navigation activity in the summer months.
- Rising sea levels may lead to saline intrusion into fresh water sources, affecting water affordability and quality.

We need to achieve more sustainable use of water despite a range of different demands (for domestic use, industry, agriculture, the natural environment, tourism, housing, etc.) both now and in the future. These demands have implications beyond the direct impacts of water scarcity and the need to reduce water use in water stressed areas while increasing the capacity of the landscape to capture and store water safely throughout the year.

The Environment Agency's abstraction licensing regime is the main mechanism for achieving sustainable management of water resources. Extensions to the scope of the licensing regime from 2009 mean that all sectors are now controlled and could have their allocations of water changed, if necessary, to respond to climate impacts. New proposals to time limit existing abstraction licences provide a further way to proactively manage the impacts of climate change on catchment areas. Defra has recently consulted on these proposals with a view to setting out a revised approach.

Because managing multiple demands for water is such a fulcrum issue, achieving and maintaining a sustainable water abstraction regime is the icon indicator for adapting to climate change under PSA27: Leading the global effort to avoid dangerous climate change. Achievement of this objective means that the overall regime for water management is adapting well to climate change.

Agriculture and Climate Change Adaptation

Defra has been working in partnership with key stakeholders including Natural England, the Environment Agency, the Forestry Commission and the Rural Climate Change Forum to identify a range of priority adaptation actions that are likely to reduce vulnerability to multiple climate risks without causing other problems for the environment and/or have multiple benefits for agricultural production, ecosystems and climate change mitigation. To respond effectively to all the threats and opportunities that climate change will bring, farmers and land managers will need to take a wide range of actions, including:

- planning, risk assessment, and monitoring; land management actions that will create important 'green infrastructure' to benefit both farms and wider society;
- changing farm types, crops and livestock;
- introducing new technology; and
- improving management of water, crops, livestock and inputs such as pesticides and fertilisers.

Many of the priority adaptation actions identified within these categories have significant co-benefits for the natural environment and are already considered best practice. Therefore their additional role in adaptation gives extra incentive for farmers to pursue them. For example, 'green infrastructure' measures such as planting trees, putting in buffer strips next to watercourses, reconnecting rivers and their flood plains, restoring or creating ponds and wetlands, are excellent examples of ways that farmers can cope with several climate change impacts that also have significant benefits for biodiversity, habitats, soil and water quality. Similarly, optimising fertiliser and pesticide inputs to increase yields, save costs and cope with new or increased pests and diseases will be highly beneficial for biodiversity and water quality.

Meanwhile potential adaptation measures that could have detrimental impacts on the natural environment have been identified – and will need to be approached in a sustainable way. It is in everyone's interests that any adaptation by agriculture enhances, or at the very least does not harm, the natural environment as healthy biodiversity and ecosystems are vital for future food production.

Many farmers say they just need good information on the impacts and possible responses in order to adapt their farm to climate change. The Agriculture and Climate Change Adaptation Project is working with industry to encourage farmers to take up the most beneficial adaptation measures, by enhancing existing advice and guidance (e.g. the industry-led Farming Futures, Government skills and advice programmes) and demonstrating best practice. National monitoring and R&D will be important to ensure that adaptation by the farming sector is sustainable and does not solve one set of problems while creating another.

Assessing Climate Risks and Building

All organisations involved in managing or advising on the natural environment should take a systematic approach to identifying climate risks and opportunities and plan how best to respond to these. For example by:

Examining how the delivery of their objectives has been affected by extreme weather events in the past. For example in addressing its responsibilities for flood risk management, the Environment Agency draws on past experience to provide support for individuals, communities and businesses to prepare for flooding. This involves following some simple steps including signing up for free flood warnings, making a flood plan and researching ways to reduce the impact of flooding on homes and properties. Considering a wide range of possible risks to their activities – including any indirect impacts, such as disruption to supply chains, or changing customer needs. For example, the Rural Climate Change Forum (RCCF) do this by bringing together industry, non-governmental organisations and Government bodies, to provide a forum for dialogue with, and advice to, Government. The forum provides advice, awareness-raising and leadership for rural stakeholders, and identify research needs on climate change and other rural land management issues.

Examining which of these are the priority risks that require an early adaptation response (these include significant risks that organisations are already facing, and those where it is important to take steps now because it could take some time to plan and implement an effective response). For example, the forestry sector has to plan ahead (fifty years or more). Adaptive measures need to be appropriate to both current and future climatic conditions so detailed planning and capacity building is required now. The Forestry Commission is working with Defra and other partners in order to build adaptive capacity and implement adaptive actions.

Identifying areas where the immediate priority is for further research and analysis to understand impacts and build a stronger basis for future decision-making. For example, in March 2009, a joint Natural Environment Research Council, Defra and DECC Ocean Acidification Research

Programme, worth £12 million, was launched. The programme will run over 5 years, and outputs will support work on adaptation.

Considering which of the risks they are dealing with could best be taken forward in conjunction with their partners, stakeholders, suppliers or customers. For example, the National Park Authorities and the Broads Authority have committed to engage with residents and visitors on the future of our National Parks to ensure we all plan for the inevitable changes ahead and to promote understanding of the adaption work planned and underway.

Improving their organisational capability through embedding adaptation risk management into their existing business planning and risk management processes and ensuring staff have access to relevant information and training. For example a range of organisations involved in land management have volunteered to take part in 'Action Research' sponsored by Defra with the aim of enhancing their capacity to address the challenge of climate change

Climate Change without Introducing Perverse Effects

We have argued that ecosystem services underpin our wider adaptive capacity and that an approach focusedon maintaining and managing healthy natural systems provides us with a framework for ensuring that adaptation actions are effective and limit the risk of perverse outcomes.

Ecosystem services can be defined as services provided by the natural environment that benefit people. These benefits include:

- resources for basic survival, such as clean air and water;
- a contribution to good physical and mental health, for example through access to green spaces, both urban and rural, and genetic resources for medicines;
- protection from hazards, through the regulation of our climate and water cycle;
- support for a strong and healthy economy, with raw materials for industry and agriculture, or through tourism and recreation;
- social, cultural and educational benefits, and wellbeing and inspiration from interaction with nature.

Through the Ecosystems Approach Action Plan (EAAP), Defra have commissioned, supported and monitored a growing body of research related to ecosystem services and their use as a framework for planning and decision making. Initiatives such as the National Ecosystem Assessment (NEA).

Five core principles underpin an ecosystems approach:

1. Taking a more holistic approach to policy-making and delivery, with the focus on maintaining healthy ecosystems and ecosystem services;
2. Ensuring that the value of ecosystem services is fully reflected in decision-making;
3. Ensuring environmental limits are respected in the context of sustainable development, taking into account ecosystem functioning;
4. Taking decisions at the appropriate spatial scale while recognising the cumulative impacts of decisions
5. Applying adaptive management of the natural environment to respond to changing pressures, including climate change.

These are supported by an approach that identifies and involves all relevant stakeholders in the decision and plan making process.

Long-term Benefits of Adaptation Actions

Understanding the economic value of the resources and services that the environment provides to society is essential if we are to make the necessary investments in maintaining and enhancing the natural systems that provide them. It will help us to make better decisions that enable society to benefit more fully from those services now, and also to avoid the potentially significant costs of having to replace these services in future should our natural systems become less effective at producing them.

New practical tools for valuation and a growing evidence base mean there is increasing scope to value environmental impacts in policy appraisal. Economic valuation can help us to understand the direct and indirect impacts of climate change on the natural environment and how this will affect society. For example, the Stern Review has supported action to reduce CO2 emissions by concluding that, if we don't act now, the overall costs and risks of climate change will be equivalent to 5-20% of global GDP each year. Valuation can also help us to assess the potential for investment in the natural environment to contribute to adaptation responses. A healthy, effectively managed natural environment delivers quantifiable economic gains for example:

- It is estimated that UK businesses could save £0.4 billion through more efficient use of water.

- The pollination of strawberries, apples, pears and other crops, is estimated to be worth around £200 million a year.
- The natural environment makes a significant contribution to the economy through tourism and recreation, with estimated expenditure on rural leisure visits amounting to £10.6bn in 2006.

The Millennium Ecosystem Assessment showed how, across the world, we are already eroding the "infrastructure" of our natural systems – our soils and seas and the communities of plants and animals that live on and in them – estimating that 60% of the services we get from our environment are being used unsustainably. The Economics of Ecosystems and Biodiversity (TEEB) project has taken this argument further, suggesting that we are losing services with a value equivalent to around 50 billion euro from land-based systems alone. These losses are equivalent to around 7% of GDP by 2050 and are forecast to intensify with climate change. Both locally and globally the financial and human costs of overstretching these resources do not necessarily accrue to those who receive the benefits and vice versa.

Making it Happen

We know that the benefits to society at large of maintaining the integrity and health of natural systems can substantially outweigh the individual (and often more immediately marketable) benefits of over-exploiting individual parts of the system. As the climate changes we will need to brigade the existing levers available to us (including market mechanisms, incentives, planning and regulation) to produce the necessary level of cooperative action needed to ensure that our environment continues to benefit individuals and society in a changing climate. Our growing knowledge of how these natural systems work together underlines the case for encouraging a strategic approach to managing land and ecosystems.

This is a challenge that will require a joined up approach across different sectors and parts of Government to ensure that we sustain and, where possible, optimise the adaptive and other benefits we receive from natural systems. This will ensure that individual policies complement one another and that integrated delivery of multiple policy objectives can be achieved, with any necessary trade-offs in the longer term made based on an analysis of the full range of costs and benefits: economic, environmental and social.

We have the knowledge and the tools to deliver a healthy natural environment and the scientific and economic arguments point to action now being both necessary and cost effective. If we work with each other and with natural systems, we will have real opportunities to increase the efficiency of delivering many of the services that society needs and values. And if we balance our demands rather than overloading natural systems, we can reduce the risk of leaving our children and future generations with higher costs for the essentials of life and a poorer planet to support them.

Adaptation, Adaptive Capacity and Vulnerability

A particular focus is on recent developments in scholarship that contribute to practical applications of adaptation and adaptive strategies. Kelly and Adger , Fussel and O'Brien et al. distinguish applications of research relating to vulnerability, including studies that relate to adaptation. The applications of interest here are those that contribute directly to adaptation initiatives to tangibly influence the vulnerability of human communities or societies to conditions related to climate change.

Adaptation in the context of human dimensions of global change usually refers to a process, action or outcome in a system (household, community, group, sector, region, country) in order for the system to better cope with, manage or adjust to some changing condition, stress, hazard, risk or opportunity. Numerous definitions of adaptation are found in climate change literature, mostly variations on a common theme. Brooks, describes adaptation as ''adjustments in a system's behavior and characteristics that enhance its ability to cope with external stress''. Smit et al., in the climate change context, refer to adaptations as ''adjustments in ecological-socio-economic systems in response to actual or expected climatic stimuli, their effects or impacts.'' Pielke , also in the climate context, defines adaptations as the ''adjustments in individual groups and institutional behavior in order to reduce society's vulnerability to climate.'' Based on their timing, adaptations can be anticipatory or reactive, and depending on their degree of spontaneity they can be autonomous or planned .

The concepts of adaptation, adaptive capacity, vulnerability, resilience, exposure and sensitivity are interrelated and have wide application to global change science. Analyses range in scale from the vulnerability and adaptation of an individual or household to a particular climate stress such as drought, through the vulnerability and adaptation of a community to multiple stresses, to the vulnerability of humankind (or the global ecosystem) to all stresses and forces. Applications also vary by the phenomena of interest (biological, economic, social, etc.), and by time scale (instantaneous, months, years, decades, centuries). While this focus includes the natural resource systems upon which societies depend, we do not review applications relating to the vulnerability and adaptation of physical or biological systems even though some of the concepts (particularly adaptation) have long, if contested, use in those fields. It is in ecological systems that the resilience concepts have been most developed. The resilience of ecosystems and socio-ecological systems is reviewed by Folke.

Practical initiatives that tangibly address and improve societal adaptive capacity, thereby reducing vulnerability, are commonly expected to be evident at the community scale. There are examples of international and national initiatives that have potential to contribute to the reduction of vulnerabilities of people, and their effects should be apparent in communities. For example, National Adaptation Plans of Action (NAPAs), if effectively implemented, should generate results evident in communities. Community is used here to mean some definable aggregation of households, interconnected in some way, and with a limited spatial extent, analgous to Coombes et al.'s use of the term ''locality.''

The following sections provide a brief overview of the concept of adaptation as it has been employed in a range of fields, and as it relates to adaptive capacity and vulnerability in the context of climate change. Then several purposes of adaptation analysis are distinguished in the climate change field, including one type of application that aims to contribute to actual adaptation strategies.

Treatment of the Adaptation Concept

The term adaptation, as it is presently used in the global change field, has its origins in natural sciences, particularly evolutionary biology. Although the definition of adaptation in the natural sciences is disputed, it broadly refers to the development of genetic or behavioral characteristics which enable organisms or systems to cope with environmental changes in order to survive and reproduce. Individual adaptations (or adaptive features) are the features of organisms which have developed to ensure survival. Consideration of adaptation within natural sciences encompasses scales from the organism or individual to the population of a single species or an entire ecosystem.

The application of the term adaptation to human systems has been traced to the anthropologist and cultural ecologist Julian Steward, who used "cultural adaptation" to describe the adjustment of "culture cores" (i.e. regional societies) to the natural environment through subsistence activities. O'Brien and Holland define the process of adaptation as "one by which groups of people add new and improved methods of coping with the environment to their cultural repertoire". Denevan considers (cultural) adaptation as a "process of change in response to a change in the physical environment or a change in internal stimuli, such as demography, economics and organization", thereby broadening the range of stresses to which human systems adapt beyond biophysical stress.

Social science treatment of adaptation in human systems has been concerned with "success" or survival of a culture. Anthropologists and archeologists suggest that adaptation is a consequence of selection acting on variation through cultural practices (adaptations) which have historically allowed a culture to survive. Cultural practices are thus equated with genetic characteristics in the natural sciences; in this Darwinian view, a group which does not have adequate methods of coping with environmental stress will not be able to compete for scarce resources and will fail to continue. In this treatment of the term, a cultural practice is an "adaptation" only if it developed to overcome stress, thereby distinguishing adaptations from "adaptive features" that allow societies to function within their environments regardless of whether or not they evolved as a result of selection.

In more recent social science work, cultural practices that allow societies to survive (and, beyond that, flourish) are considered adaptations which can be distinguished based on behavior and (technological) innovation. It is recognized that societies adapt to a range of stimuli including, but not limited to, environmental stress. Cultures (or societies) which are able to respond to or cope with change quickly and easily are considered to have high "adaptability" or "capacity to adapt".

The concept of adaptation has been used both explicitly and implicitly in the social sciences, including in natural hazards, political ecology, and the entitlements and food security scholarship. Some scholars of adaptation have employed the concepts and terminology of biophysical ecological change with a focus on flows of matter, energy and information and related concepts of resilience, equilibrium and adaptive management. Others, particularly in the natural hazards perspective, have focused on perception, adjustment and management of environmental hazards.

Adaptation is usually implicit in the political ecology field. The relationships between ecosystems and political economy are often treated as issues of adaptive management of risks related to political and social power relations, resource use, and global economies. Work on entitlements and food security considers adaptation as a stress response in light of access to resources and the abilities of people to cope. A key feature of this field is its demonstration of how the adaptive capacity of individuals or households is shaped and constrained by social, political, and economic processes at higher scales. Similarly, research on global environmental risk and the social amplification of risk places adjustments and adaptations in the context of human driving forces, biophysical constraints and the social, economic and political attenuation of risks.

Conceptualizations of risks and their manifestation as disasters, including the pressure and release (PAR) mode, identify the environmental stresses of hazards and the progression of social forces that contribute to vulnerability, including those that relate to adaptive capacity. This view of environment–society coupled systems that specify the role of human adaptive responses is further developed in the vulnerability framework of Turner et al. and the access model of Wisner et al..

Analyses of adaptations in the climate change field emerged concurrently with the growing awareness of climate change itself. An early example is Butzer who considered ''cultural adaptation'' (human ingenuity including technological innovation and long-range planning) in light of predicted climate change and its anticipated impacts on world food supply. Since then, analyses of adaptation to changing climatic conditions have been undertaken for a variety of purposes.

Climate Change Adaptation Research

One common purpose of adaptation analyses in the climate change field is to estimate the degree to which modeled impacts of climate change scenarios could be moderated or offset (or ''mitigated'') by ''adaptation to the impacts'. These analyses address Article 2 of the United Nations Framework Convention on Climate Change (UNFCCC), which commits countries to mitigate greenhouse emissions in order to avoid ''dangerous'' anthropogenic changes in climate. Adaptations are considered to assess the degree to which they can moderate or reduce negative impacts of climate change, or realize positive effects, to avoid the danger. These analyses are usually undertaken at broad scales, where equilibrium or statistical models are used to estimate impacts with and without adaptation, in order to address the question: how serious or "dangerous" are specified scenarios of climate change.

When analyzed for this purpose, adaptations are conventionally assumed or hypothetical, and their effect on the system of interest is estimated relative to the estimated impacts (e.g. in terms of costs, savings, etc.). For this use, the focus is on the effect of the assumed adaptations. The purpose is to estimate impacts of climate change, and to estimate the difference adaptation could make. This work does not empirically investigate adaptations, examine the actual processes of adaptation or adaptive capacity, explore the conditions or drivers that facilitate or constrain adaptations, nor document the decision-making processes, authorities and mechanisms involved in adaptation. It takes certain assumed or hypothetical adaptations and then estimates the effects they would have on the calculated impacts of conditions captured in the specified climate change scenarios. The term vulnerability has sometimes been used to describe the estimated net or residual impacts (initial impact costs minus net adaptation savings).

A second body of scholarship focuses on specific adaptation options or measures, for a particular system subject to climate change stimuli. These analyses address the articles of UNFCCC that commit countries to "formulate and implementy measures to facilitate adequate adaptation to climate change" (Article 4.1). The purpose of these analyses is to assess the relative merit or utility of alternative adaptations, in order to identify the "best" or better ones. The analysis involves selecting a suite of "possible adaptations", chosen by the researcher from hypotheses, observations, modeling, extrapolation, analysis, key informants or deductive reasoning. These possible adaptations are usually considered to be distinct and discrete, in order that they can be subjected to evaluation according to some common principles or criteria. Among the tools used to rank or rate the relative merit of possible adaptations are benefit-cost, cost effectiveness and multiple-criteria procedures. Common variables employed are benefits, costs, implementability, effectiveness, efficiency, and equity . Such analyses assume that there is, in practice, a process through which adaptations are selected and implemented, and that the relative evaluation analysis fits into this process. The focus of these studies is to rate or rank potential adaptations, but they rarely investigate the processes through which adaptation measures are undertaken, either in light of climatic change specifically (which is very rare) or as part of policy and decision-making processes to which adaptations to climate change might relate.

A third group of studies focuses on the relative adaptive capacity (or vulnerability) of countries, regions or communities, and involves comparative evaluation or rating based on criteria, indices and variables typically selected by the researcher. Vulnerability is taken as the ''starting point'' rather than the residual or ''end point'', and it is assumed to be measurable based on attributes or determinants selected a priori. The expected application is that adaptation efforts should be directed to those areas with the greatest exposures or least adaptive capacity.

This work relates to UNFCCC Article 4.4, which commits developed country parties to ''assist developing country parties that are particularly vulnerable to the adverse effects of climate changey''. The main purpose of these studies is to provide an evaluation of the relative vulnerability (and/or relative adaptive capacity) of the countries or regions, usually using some kind of indicator, scoring, rating or ranking procedure. Thus, surrogate measures of exposure or sensitivity and elements of adaptive capacity for each system are estimated and then aggregated to generate an overall vulnerability ''score'' (or level or rating) for each system. The intent is to provide information for the targeting of adaptation initiatives, or the targeting of scarce resources.

In this third type of research, the analyst selects the factors or determinants of vulnerability or adaptive capacity (sometimes with local inputs), obtains measures on these (usually aggregate surrogates from available secondary data), adopts an aggregation function over the measures (usually summation) and calculates an overall vulnerability value for each system. This research does not aim to identify the processes, determinants or drivers of adaptive capacity and vulnerability as they function in each system—they are taken as given, and used as the basis for the rating or ranking analysis. Nor does this analysis substantively address the policy and decision-making processes that deal with the conditions that can alter adaptive capacity and vulnerability. It is implicitly assumed that the output—indications of the relative vulnerability or adaptive capacity—will have application in policy and decision-making, by identifying the countries or districts or areas with the greatest vulnerability or least adaptive capacity.

The purpose of the fourth type of analysis is to contribute to practical adaptation initiatives. Research that focuses on the implementation processes for adaptations is still not common; at least, it is not common under the label of "adaptation" research, and certainly not in the climate change field. There is a vast body of scholarship in the fields of resource management, community development, risk management, planning, food security, livelihood security, and sustainable development that deals with the actual practices and processes of adaptation, although the word "adaptation" may not be explicitly used.

By "practical application", we mean research that investigates the adaptive capacity and adaptive needs in a particular region or community in order to identify means of implementing adaptation initiatives or enhancing adaptive capacity. This enables the identification and development of particular adaptive measures or practices tailored to the needs of that community. The aim is not to score adaptations or measure relative vulnerabilities, nor to quantify impacts or estimate effects of assumed adaptations. Rather, the focus is to document the ways in which the system or community experiences changing conditions and the processes of decision-making in this system (or that influence the system) that may accommodate adaptations or provide means of improving adaptive capacity.

In the climate adaptation field, this body of work is characterized by several distinctive features that are important to facilitate adaptation initiatives. It tends not to presume the specific variables that represent exposures, sensitivities, or aspects of adaptive capacity, but seeks to identify these empirically from the community. It focuses on conditions that are important to the community rather than those assumed by the researcher or for which data are readily available. It employs the experience and knowledge of community members to characterize pertinent conditions, community sensitivities, adaptive strategies, and decision-making process related to adaptive capacity or resilience. It identifies and documents the decision-making processes into which adaptations to climate change can be integrated. It is sometimes called a "bottom-up" approach in contrast to the scenario-based "top-down" approaches.

The distinctive motivation here is to identify what can be done in a practical sense, in what way and by whom, in order to moderate the vulnerability to the conditions that are problematic for the community. This work is not designed to be "scaled up" in the sense of generating an aggregate regional or global score or valuation of vulnerability or adaptation. The "scaling up" of this work would involve comparisons across communities or societies in order to identify those characteristics of communities and their environments that contribute to or moderate vulnerabilities, and the features of adaptive strategies that are effective.

One noteworthy development in this applied work on adaptation processes is that of "mainstreaming". The whole point of the work on adaptation processes is to have risks (and opportunities) associated with climate change (or other environmental changes) actually addressed in decision-making at some practical level. One of the fundamental findings from this work is that it is extremely unlikely for any type of adaptive action to be taken in light of climate change alone.

There are now numerous examples of climate change risks being incorporated into existing policies, programs or decision-making processes related to resource management, community development, livelihood enhancements, coastal zone management, sustainable development and risk management. Practical climate change adaptation initiatives are invariably integrated with other programs, and often aim to enhance adaptive capacity.

Adaptive Capacity and Vulnerability

Adaptation, whether analyzed for purposes of assessment or practice, is intimately associated with the concepts of vulnerability and adaptive capacity. A general conceptual model of vulnerability has emerged in the climate change scholarship, similar to the use of the concept more widely. Consistent throughout the literature is the notion that the vulnerability of any system (at any scale) is reflective of (or a function of) the exposure and sensitivity of that system to hazardous conditions and the ability or capacity or resilience of the system to cope, adapt or recover from the effects of those conditions. These concepts are labeled in different ways and given different emphases in various fields. Adaptations are manifestations of adaptive capacity, and they represent ways of reducing vulnerability.

The larger sets represent the broader stresses and forces that determine exposure and sensitivity and shape adaptive capacity at the local or community level, denoted by the smaller embedded sets. The interaction of environmental and social forces determines exposures and sensitivities, and various social, cultural, political and economic forces shape adaptive capacity. The overlap recognizes that the processes driving exposure, sensitivity and adaptive capacity are frequently interdependent. The finer scale interaction of these elements represents local vulnerability, and adaptations are particular expressions of the inherent adaptive capacity. Generally, a system (e.g. a community) that is more exposed and sensitive to a climate stimulus, condition or hazard will be more vulnerable, Ceteris paribus, and a system that has more adaptive capacity will tend to be less vulnerable, Ceteris paribus.

This conceptualization broadly indicates the ways in which vulnerabilities of communities are shaped. It does not necessarily imply that the elements of exposure, sensitivity and adaptive capacity can or should be measured in order to numerically compare the relative vulnerability of communities, regions or countries. Vulnerability, its elements of exposure, sensitivity and adaptive capacity, and their determinants are dynamic (they vary over time), they vary by type, they vary from stimulus to stimulus, and they are place- and system-specific.

Consistent with the literature, this conceptualization differentiates the two broad contributing elements of vulnerability, but does not suggest that these are unrelated. The model does not specify a priori particular factors, processes or functional relationships between exposure, sensitivity and adaptive capacity. It assumes they exist and are distinctive to particular places and times. Certainly, there are broad social, economic, political and ecological conditions that affect exposure, sensitivity and adaptive capacity, but at the community level these elements will be exhibited in diverse ways. Even for a particular system, vulnerability is unlikely to be the same for all stimuli, even all climatic stimuli (e.g. increasing temperature, floods, sea level rise, low-frequency droughts, high-frequency droughts, high-frequency extended droughts, etc.).

Exposure and sensitivity are almost inseparable properties of a system (or community) and are dependent on the interaction between the characteristics of the system and on the attributes of the climate stimulus. The exposure and sensitivity of a system (e.g. a community) to an environmental change risk (e.g. drought) reflect the likelihood of the system experiencing the particular conditions and the occupance and livelihood characteristics of the system which influence its sensitivity to such exposure. The occupance characteristics (e.g. settlement location and types, livelihoods, land uses, etc.), reflect broader social, economic, cultural, political and environmental conditions, sometimes called "drivers" or "sources" or "determinants" of exposure and sensitivity. Many of the determinants of occupance or sensitivity are similar to those that influence or constrain a system's adaptive capacity. A research challenge for practical adaptation work to characterize the exposure and sensitivity elements of vulnerability is to identify those processes of climate conditions and system occupancy dynamics that are considered to be problematic, risky or hazardous in some way to the community of interest. These are rarely known a priori.

Adaptive capacity is similar to or closely related to a host of other commonly used concepts, including adaptability, coping ability, management capacity, stability, robustness,flexibility, and resilience. The forces that influence the ability of the system to adapt are the drivers or determinants of adaptive capacity. Local adaptive capacity is reflective of broader conditions. At the local level the ability to undertake adaptations can be influenced by such factors as managerial ability, access to financial, technological and information resources, infrastructure, the institutional environment within which adaptations occur, political influence, kinship networks, etc. Some determinants of adaptive capacity are mainly local (e.g. the presence of a strong kinship network which will absorb stress) while others reflect more general socio-economic and political systems (e.g. the availability of state-subsidized crop insurance).

Adaptive capacity is context-specific and varies from country to country, from community to community, among social groups and individuals, and over time. It varies not only in terms of its value but also according to its nature. The scales of adaptive capacity are not independent or separate: the capacity of a household to cope with climate risks depends to some degree on the enabling environment of the community, and the adaptive capacity of the community is reflective of the resources and processes of the region.

Adaptive capacity has been analyzed in various ways, including via thresholds and "coping ranges", defined by the conditions that a system can deal with, accommodate, adapt to, and recover from. Most communities and sectors can cope with (or adapt to) normal climatic conditions and moderate deviations from the norm, but exposures involving extreme events that may lie outside the coping range, or may exceed the adaptive capacity of the community. Some authors apply "coping ability" to shorter term capacity or the ability to just survive, and employ "adaptive capacity" for longer term or more sustainable adjustments. Watts and Bohle use "adaptability" for the shorter term coping and "potentiality" for the longer term capacity.

A system's adaptive capacity and coping range (one feature of capacity) are not static. Coping ranges are flexible and respond to changes in economic, social, political and institutional conditions over time. For instance, population pressure or resource depletion may gradually reduce a system's coping ability and narrow its coping range, while economic growth or improvements in technology or institutions may lead to an increase in adaptive capacity.

The graphical representation shows that the coping range(in this case to deal with drought) can increase over time or decrease, for a variety of reasons. External socio-economic and political factors (e.g. war, the collapse of an institution such as a crop insurance program, loss of a key decision-maker) may lead to a narrower coping range. Furthermore, the cumulative effects of increased frequency of events near the limit of the coping range may decrease the threshold beyond which the system cannot cope/adapt/recover. For example, two consecutive years of high moisture deficit which are not beyond the limits of the normal coping range present little problem in the present but require drawing on stored resources, and the consumption of these resources may subsequently narrow the coping range until they can be built up again, so a third and fourth year of moisture deficit of the same magnitude may well exceed the now smaller coping range.

Similarly, conditions which are within the coping range may introduce unforeseen side effects which will narrow the coping range. For example, a warm, wet year may be an ideal year for crop production and lead to high yields. Subsequent years of warm, wet conditions can, however, encourage the development of pest and fungal outbreaks and actually decrease yields and thus the coping range is reduced. Finally, a catastrophic event beyond the limit of the coping range may permanently alter the system's normal coping range if it is not able to recover from it. For example, consider a system that relies on irrigation water, captured in a dam. A very wet year, far beyond the normal conditions, may lead to the dam's failure, and thus the previous coping range cannot be returned to in a subsequent "average" year.

Adaptations are manifestations of adaptive capacity. Adaptations, or changes in the system to better deal with problematic exposures and sensitivities, reflect adaptive capacity. Clearly there are many forms and "levels" of adaptations, and these can be classified in many ways including by timing relative to stimulus (anticipatory, concurrent, reactive), intent (autonomous, planned), spatial scope (local, widespread) and form (technological, behavioral, financial, institutional, informational. It is also possible to distinguish adaptations according to the degree of adjustment or change required from (or to) the original system. For an agricultural system facing water shortage exposures, a simple adaptation might be to use more drought resistant cultivars. A more substantial adaptation might be to shift away from crop farming to pastoralism. An even more substantial adaptation might be to abandon farming altogether.

The determinants of adaptive capacity are not independent of each other. For example, the presence of a strong kinship network may increase adaptive capacity by allowing greater access to economic resources, increasing managerial ability, supplying supplementary labor and buffering psychological stress. Similarly, economic resources may facilitate the implementation of a new technology and ensure access to training opportunities and may even lead to greater political influence. Individual determinants, thus, cannot be isolated: adaptive capacity is generated by the interaction of determinants which vary in space and time. The determinants of adaptive capacity exist and function differently in different contexts. For example, a strong kinship network may play an important role in a subsistence-based society, and quite a different role in a developed world agribusiness context where financial and institutional structures will influence adaptability.

To date, there is very little consensus (or documented support) for a robust, specific model of the elements and processes of local exposure, sensitivity, and adaptive capacity, beyond broad categories. The broad factors or determinants that influence sensitivities and constrain the abilities of communities to deal with hazards or stressful conditions are too general guide in practical adaptation programs. Community-based analyses have shown that the conditions that interact to shape exposures, sensitivities, adaptive capacities, and hence create needs and opportunities for adaptation, are community specific. For example, the factor "technology" may be relevant in all cases, but the way in which technologies influence vulnerabilities and the types of technologies that may be feasible or available and how they interact with political, social and economic processes invariably differ from community to community

Adaptation Analysis to Practice

Some general principles are now apparent from community-based vulnerability assessments aiming to contribute to practical adaptation initiatives. One is that the researcher does not presume to know the exposure and sensitivities that are pertinent to the community, nor does the research specify a priori determinants of adaptive capacity in the community. Rather, in this approach these are identified from the community itself. The methods require the active involvement of stakeholders, considerable effort to ensure legitimacy, information collection on community relevant phenomena and processes, the integration of information from multiple sources, and the engagement of decision-makers.

Variants of participatory, ''bottom-up'', experience-based assessment of community conditions have been employed in many fields including sociology, anthropology, geography, ethnography, risk assessment, rural development, international development and food security. In the climate change adaptation and disaster management fields, analytical frameworks very similar to these have been developed and some have been applied.

Participatory vulnerability assessments allow for the recognition of multiple stimuli beyond those related to climate, to include political, cultural, economic, institutional and technological forces. Furthermore, the methodologies recognize the interaction of various exposures, sensitivities and adaptive capacities over time. What is vulnerable in one period is not necessarily vulnerable (or vulnerable in the same way) in the next, and some exposures and sensitivities develop slowly over time. Finally, the approach recognizes that sources of exposures, sensitivities and adaptive capacities function across scales, from the individual to the national.

The system of interest in this case is the community, but the analysis seeks to identify the broader conditions and structures within which the community functions. The exercise requires active involvement of community stakeholders. Researchers begin with an assessment of current exposures, sensitivities and current adaptive capacity, employing ethnographic in-community methods (including such tools as semi-structured interviews, participant observation and focus groups), as well as insights from local and regional decision-makers, resource managers, scientists, published and unpublished literature, and other available sources of information.

Once relevant conditions have been identified, and future livelihoods considered, information from other scientists, policy analysts, and decision-makers, are integrated into the analysis to identify potential future exposures and sensitivities (what conditions or risks the community may be facing) and future adaptive capacity (in what ways the community may potentially plan for or respond to these conditions) to determine future vulnerability. Opportunities to reduce future vulnerabilities are sought with community decision-makers, and representatives of other agencies with authority or influence. Experience to date has shown that the common adaptation practices involve modifying some existing resource management strategy, livelihood enhancement initiatives (e.g. income diversification in Bangladesh), disaster preparedness plan (e.g. flood or hurricane warning and planning in coastal Vietnam), or sustainable development program.

The goal of the methodology outlined above is not to produce a score or rating of a particular community's current or future vulnerability. Rather, the aim is to attain information on the nature of vulnerability and its components and determinants, in order to identify ways in which the adaptive capacity can be increased and exposure-sensitivities decreased. While adaptation options are evaluated in some way, the initiatives are rarely discrete stand-alone, exclusively climate change measures that are amenable to comparative scoring. Instead, adaptation initiatives tend to be incremental, modifying some existing water management strategy, disaster plan, and so on. This is commonly known as mainstreaming. Successful climate change adaptation and vulnerability reduction is rarely undertaken with respect to climate change alone, and vulnerability reduction appears to be most effective if undertaken in combination with other strategies and plans at various levels.

Adaptation is still a novel concept to some in the climate change field, but is has considerable history in others fields. That work has shown that adaptations in human communities are closely associated with, and reflective of, adaptive capacity and vulnerability. In particular, it has shown that vulnerability is related both to the differential exposure and sensitivity of communities to stimuli such as climate change and also to the particular adaptive capacities of those communities to deal with the effects or risks associated with the exposures. While exposures, sensitivities and adaptive capacities are evident at community or local levels, they reflect broader forces, drivers or determinants that shape or influence local level vulnerabilities.

Studies of adaptation to climate change have provided many insights but to date, have shown only moderate practical effect in reducing vulnerabilities of people to risks associated with climate change. One widely acknowledged lesson is that adaptations are rarely undertaken in response to climate change effects alone, and certainly not to climatic variables that may be of importance to decision-makers. The broader literature has also shown the utility of including decision processes in the exercise if it is aiming to affect implementation. There has been considerable scholarship in the climate change context on calculating indices of vulnerability and adaptive capacities, and on evaluating hypothetical adaptations, yet the practical applications of this work (in reducing vulnerabilities of people) are not yet readily apparent. Some success in practical implementation has been seen when measures that address climate change risks are incorporated into existing decisions structures relating to risk management, land use planning, livelihood enhancements, water and other resource management systems, development initiatives, and so on.

This movement to ''mainstreaming'' adaptation to climate change is consistent with the broader literature on how adaptations and adaptive capacity work. That literature also makes it clear that local initiatives, to enhance livelihoods and hence adaptive capacity, may be constrained or even nullified by broader social, economic and political forces that effectively shape local vulnerabilities. This brings the adaptation issue to the question of development and the role of local initiatives relative to transformations of geo-political–economic systems.

References

Linacre, E., (1992). *Cliamte data and resources: a reference and guide.* Rouledge, London, 366pp.

Mitchell, J.M. *et al.*, (1966), *Climate Change. Technical Note No. 79.* World Meteorological Organisation.

Trewartha, G.T. & Horn, L.H., (1980). *An Introduction to Climate.* New York: McGraw-Hill.

United Nations Environmental Programme / World Meteorological Organisation (UNEP/WMO), (1992). *Framework Convention on Climate Change.* United Nations, Geneva.

Wigley, T.M.L. & Barnett, T.P., (1990). Detection of the Greenhouse Effect in the Observations. In: *Climate Change: The IPCC Scientific Assessment,* Intergovernmental Panel on Climate Change (IPCC), 1990.

5

Linking Economy, Society and Environment

Two thousand three hundred miles to the west of Chile and 1 300 miles to the east of Polynesia's Pitcairn Islands lies an island that has inspired intense interest for centuries, not for its perfect climate or its untouched beauty, but because it holds a secret, a mystery. Rapa Nui, or Easter Island as it was named by 18th century Dutch explorers, attracts scientists from around the world who come to study its stone statues, called moai.

The moai, like the pyramids of ancient Egypt, intrigue and confound us with their sheer size: weighing up to 270 tonnes and as much as 70 feet tall, these massive monolithic figures form an imposing presence: outsized human faces looking out over this remote island and the thousands of miles of ocean that lie beyond it. We marvel at the engineering and wonder at how stone-age Polynesians managed to erect such immense structures without the use of cranes, metal tools, or large animals. The creation of statuary of this size and sophistication speaks to the existence of a populous, creative and complex society – one that was well-off enough to support an artisan class. They could afford to allocate time and resources to the various activities involved in making, transporting and erecting hundreds of statues. Or could they? European explorers who visited the island in the 18th and 1 9th centuries f ound a population of only a f ew thousand, a mere remnant of the statue-building society that came before. Something had significantly altered life on Rapa Nui.

What had at one time been a sub-tropical forest was now a completely deforested island, with at least 22 species of trees and plants extinct. Most wild sources of food were gone – overhunting had left Easter with almost no wild bird species. Without trees to make canoes, large fish were inaccessible, leaving only fish that could be caught close to shore. Evidence shows that these stocks too were depleted. What happened to bring Easter's civilisation to near extinction, driving its population almost to zero and ending its period of cultural flourishing and creative production? Forest was cleared for agriculture. Trees were also burned to obtain charcoal.

Another possibility is that rats brought to the island by the first settlers fed off the seeds of the trees. Easter's collapse has inspired thousands of pages of study and analysis – in part because islands make interesting cases studies, providing a kind of closed P etri dish in which we can study cause and effect. But Easter also intrigues us because of the extent of its devastation, what Diamond calls "the most extreme example of forest destruction in the Pacific, and among the most extreme in the world." Is there a lesson in this experience for the world of today? What can we learn from Easter's cautionary tale?

The relationship of humans to the environment has always been one of give and take. Easter Islanders made use of their surroundings for their physical and cultural needs in the same way that all human societies do – but they either did not see or did not heed the requirement of keeping their "systems" in balance, of ensuring that new trees were growing when old trees were cut, for example. When the rate of use overtakes the rate at which a resource can be replenished, then that resource will be drawn down and eventually disappear, affecting all of the people, animals and plants that depend on it. The question of equilibrium – balancing use with renewal, pollution with its impact on ecosystems – is key to understanding the challenges of our world. Even CO_2 emissions that we all worry about these days serve a beneficial purpose, being absorbed by plants for increased growth, as long as the proportions remain right: the carbon dioxide put into the atmosphere should not exceed what can be absorbed through photosynthesis. Problems arise when proportions get out of balance, such as with excessive CO_2 emissions that cannot be absorbed by the ocean, plants and other so-called carbon sinks, and thus contribute to climate change.

Keeping systems in balance is an important idea that reaches beyond environmental concerns. Think of the demographic balances in a given society, the interplay between births, deaths, emigration and immigration. For our economies, we must have enough young workers to replace retirees, and to fund their pensions. Finally, can societies remain stable when resources are concentrated in the hands of a few, while others go without?

World today

To look at the statistics, the world today is, on average, a prosperous place. Growth in the second half of the 20th century was greater than at any previous historical period. Average incomes have increased eightfold since 1820, while population is five times higher.

The world economy performed better in the last half century than at any time in the past. World GDP increased six-fold from 1950 to 1998 with an average growth of 3.9 per cent a year compared with 1.6 from 1 820 to 1950, and 0.3 per cent from 1500 to 1820.

Global life expectancy at birth in 1800 was about 30 years, compared with 67 in 2000 and 75 in the rich countries. In countries with well-developed health care systems, infant mortality has been brought to very low levels and vaccines have virtually eliminated life-threatening childhood diseases.

We also live in a period of intense cultural production and technical ability. The so-called information age has put virtually limitless amounts of data at our fingertips - provided that we have access to the technology that links us to it. Films, plays, books, music, scientific studies, analysis and opinion on everything from politics to sport are all readily available, creating the possibilities for a society that is better informed and more aware than in any previous historical epoch.

And we are not just learning or consuming all of this content as individuals – we are discussing it, interacting with it and refining it collaboratively. Blogs, wikis, website discussion threads: these have created a new nexus of information between "official" and "unofficial" communications. Some bloggers become authorities on their topics and influence trends. Wiki contributors become widely read. The lines of communication have essentially opened up, giving us the opportunity and the responsibility of understanding what is going on around us – provided that we learn to use all of this information in a meaningful way.

Indeed, our choices have multiplied in nearly every domain: educational, professional and personal. As students we can choose from hundreds of subjects of study and among an increasing number of educational institutions offering diplomas. Programmes like the EU's Erasmus exchange scheme encourage students from one country to study in another – to learn another language, another culture, or simply to have access to a particular type of education not available in their home country.

The globalisation of business, science and culture has also opened up our professional choices: following a job far from one's home town, working as an expatriate in another country, travelling regularly to offices around the world. On the whole, we are a wealthier, longer living, more educated and more mobile population. But can this continue? Will it be true for future generations? In all parts of the world?

Clouds on the Horizon

Still, in spite of the advanced state of many contemporary societies, we see some troubling contradictions. Notably, there is a stark inequality between those with access to the fruits of advanced development, and those living in contexts where that advancement is impeded by lack of access to what others take for granted.

Stark differences divide the world in terms of access to water and sanitation, energy, health care and education. For example, it is estimated that 1.1 billion people in the world lack clean water. The question is not one of comfort: water-related illnesses are the second biggest killer of children in the world – approximately 1.8 million children die each year from diseases caused by dirty water and poor sanitation. Illness from poor water and sanitation keeps children out of school and adults out of work, while the search for water in areas where access is poor takes up a large portion of time in the daily lives of women and girls, time that they cannot spend working for economic improvement or going to school.

According to the United Nations Human Development report and to water specialists like Professor A.K. Biswas, the problem is not one of scarcity, but mismanagement. Leaking taps in the developed world waste more water than is available to the billion people in the developing world who need it. Fixing those leaking taps won't magically solve water access problems, but an approach to water management that includes sharing successful techniques for making the best use of available water supplies can improve things dramatically.

While people living in the least developed nations often lack the necessary elements to fulfil basic needs and have access to a life of health and quality, the developed world suffers from having too much. Poorer countries face the terrible consequences of largely preventable diseases like malaria or AIDS, while the richer ones battles epidemics of excess, such as adult-onset (or Type II) diabetes and heart disease caused by obesity. There is a level of international co-operation never before seen in history bilaterally, or between governments, and multilaterally through institutions like the United Nations, OECD, World Bank and others. And yet there are still violent conflicts that place those caught in them in conditions of extreme insecurity and vulnerability. The human population continues to grow. Predictions have the current population of 6.5 billion increasing to over 8 billion by 2050. More and more of those people are living in cities, and everyone who has the means is using more resources. Our lives are full of more and more things. The proliferation of markets, products, and the ease of trade means that both our choices for consumption and the consequences of that increased activity are greater than ever.

Economic development has allowed for advances that have fundamentally changed the ways humans live from previous centuries, but these activities have also brought about problems with potentially dramatic consequences. Climate change is the most visible, most talked about at the moment, especially after the recent Intergovernmental Panel on Climate Change report confirmed that the climate is almost certainly undergoing significant change as a result of human activity. But economic development has also brought social challenges: countries are advancing at different speeds, and people within countries are living with vastly different quality of life. In many countries the gap between the rich and the poor is increasing rather than shrinking with economic growth.

Climate change is symbolic of the larger problem – one that is both practical and philosophical – of the dangers inherent in pushing our ecosystems out of balance. Are we pushing our societies and environment too far, too fast? Are we outrunning the regenerative possibilities inherent to our ecosystems? Are we creating social imbalances that cannot be corrected? Could we be on some kind of crash course, like the Easter Islanders, without even realising it? When systems work, when they are in equilibrium, they tend to continually produce possibilities for renewal: if land is well-managed, given time to lie fallow and regenerate nutrients, then it continues to be fertile indefinitely. If not, then the quality of soil degrades and in some cases becomes useless. Wild species naturally replace themselves. But populations will crash, possibly to extinction, if they are overharvested.

We can even extend this n otion to h umans and their interactions. Children who are well-nourished, educated and cared for tend to flourish, carrying with them a lifelong capacity to contribute to their community. Deprive them of those things and the outcome is likely to be quite different. The same is true at the societal and governmental l evel. Abuse, conflict or deprivation can cause entire communities to collapse. This can be applied to economic systems or markets. Imbalances in supply and demand, in savings and s pending, in l oans and investment can lead to economic crashes, recessions and depressions. The most talented economists are still unable to predict reliably when and why these events might occur, due to the extreme complexity of the world's economy. W hat w e do know is t hat economic, environmental and social systems must all be kept in relative equilibrium, and also balanced with each other, to be sustainable. One problem is that we do not know w hen the "critical threshold" of these systems will be reached and exceeded. To continue moving towards, even beyond these thresholds, is to take a great risk: are we creating a future that will experience failures of Earth's life-sustaining systems with increasing frequency and unpredictability? Are we living in a present where economic and social developments benefit some and leave others mired in need and conflict?

World's Economy

In the last 200 years, the world's economy has grown sixfold, and almost tenfold in the regions that were first to industrialise. Standards of living, health and education have improved considerably. At the same time, burning coal for energy led to deadly smog in England and the US, water pollution left entire lakes "dead", irrigation for cotton brought the Aral Sea to a fraction of its historical area, and now fossil fuel use is causing changes to our climate. In addition, economic and technological development has left huge gaps in prosperity, opportunity and standards of living. The question is: can we do a better job with development, starting now?

What are the principles driving these phenomena? What kind of future are they spelling out for our descendents? As we develop economically and socially, whether as individuals, governments or businesses, we need guiding principles that will help us make the right choices. But do we really have to choose between progress and sound management of the systems that support us? Every day, we hear about new technologies that can benefit people, economy *and* the environment: public health programmes that improve health outcomes for more people, energy efficient alternatives for many of the products and processes we have come to rely on, and new non-toxic and durable materials.

If the Easter Islanders were aware of their dwindling resource base, history shows that they didn't take the necessary steps to prevent passing the critical threshold. Many people today realise that our world also shows signs of stress – and at the very least presents some core problems to which we need to find solutions. Evidence suggests that we need better ways of managing our natural resources; better ways of securing what people need to develop; better ways of co-ordinating our actions to take care of all the things we rely on to survive, thrive and prosper. It is time to learn how to develop without these negative social and environmental side effects, and in a way that benefits more of us. Easter Island was isolated from trade and limited in its ecological resources; perhaps the only way its human residents could have prevented tragedy was through careful planning. We are living on a much larger scale, but could the same be true of us?

Defining Sustainable Development

Monique Huteau, a recently retired nurse, is up to her elbows in soil as she tends to her lush garden. Strawberries, lettuce, squash, potatoes and more, she grows enough to cover a large portion of her family's needs in produce, buying what she lacks at the local hypermart. She also cooks, cleans, cares for her grandchildren and paints watercolours at her home in the countryside a few minutes outside Poitiers, France.

During their working years, she and most of her nine siblings earned considerably more than their parents had, poor farmers from the Anjou region. They live in well-maintained houses, drive nice cars, and take yearly vacations to distant places. A lot of hard work and astute savings and investments have allowed Monique and her husband, a retired teacher, to achieve these things – with the help of French social benefits which have kept their health care and education costs low and ensured them an adequate retirement income. For Monique there is no question: her generation had opportunities not available to their parents and consequently live very different lives materially and socially. Monique's experience is a common one in OECD countries: generations that had endured scarcity and hardship growing up have achieved, even on relatively modest incomes, the satisfaction of basic needs plus enough discretionary income to indulge in a few luxuries. With this have also come certain social benefits. Education levels have increased. More people have access to health care. Leisure time is guaranteed through paid vacations and retirement plans. There is more geographic and social mobility.

Indeed, the so-called developed world has seen average improvements in many areas that are important to "the good life". Along with these improvements, however, have come worrying indications that this growth has costs which we cannot continue to ignore. All the economic prosperity in the world cannot alone solve a problem like climate change on its own. On the contrary , unchecked growth – in the n umber of people driving cars and taking planes, for example – is making the situation worse. Also, average economic growth says nothing about income inequality: if wealth is growing for only a few, then the majority may not experience any tangible gains or improvements.

These problems are compounded when added to the challenges facing the developing world - encompassing countries such as China and India who are undergoing rapid growth, as well as those like many Sub-Saharan African countries which are still far from having what the richer countries take for granted: peace, basic health care access, education, a relatively safe water supply, and so on.

Finally, do the resources exist to enable this kind of lifestyle for all of Earth's 6.5 billion residents? It is estimated that in 2002 humans extracted over 50 billion tons of natural resources from the planet's ecosystems, up a third in just 20 years. Projected economic growth rates put our extraction needs at 80 billion tons in 2020. Would using the Earth's resources at this rate be advisable? Can we and should we continue with the traditional model of development? These problems are not new. Indeed, the accumulation of a number of bad habits and "unsustainable" practices seems to have led to critical stresses on societies and the environment. In spite of unprecedented economic growth, the world has been on a course leading to resource depletion and serious social crises, and old ways of problem-solving have proven inadequate. Something has to be done to change development - its philosophy and methods -if societies wanted to reverse those negative trends.

As Albert Einstein wrote, “Today’s problems cannot be solved if we still think the way we thought when we created them.”

The term sustainable development began to gain wide acceptance in the late 1980s, after its appearance in *Our Common Future,* also known as *The Brundtland Report.* The result of a UN-convened commission created to propose “a global agenda for change” in the concept and practices of development, the Brundtland report signalled the urgency of re-thinking our ways of living and governing. To “responsibly meet humanity’s goals and aspirations” would require new ways of considering old problems as well as international co-operation and co-ordination.

The World Commission on Environment and Development, as it was formally called, sought to draw the world’s attention to “the accelerating deterioration of the human environment and natural resources and the consequences of that deterioration for economic and social development.” In establishing the commission, the UN General Assembly explicitly called attention to two important ideas:

- The well-being of the environment, of economies and of people is inextricably linked.
- Sustainable development involves co-operation on a global scale.

Sustainable development is about integration: developing in a way that benefits the widest possible range of sectors, across borders and even between generations. In other words, our decisions should take into consideration potential impact on society, the environment and the economy, while keeping in mind that: our actions will have impacts elsewhere and our actions will have an impact on the future.

We tend to arrange things compartmentally, by divisions and departments, governments and communities; even households are rarely set up as holistic systems. Ministries of agriculture, finance, the interior and foreign affairs handle the issues that come under their domain. We divide up the tasks of our daily lives: work, rest, errands and holidays. It is not that we *can’t* see business, government or home life as a “whole” – making a household budget or a corporate strategy are examples of just this type of exercise – but in the bustle of our complex lives it can be difficult to take the time to see beyond the most immediate or obvious concerns. Often, as the old saying goes, we can’t see the forest for the trees.

The concept of sustainable development has been used to articulate several essential shifts of perspective in how we relate to the world around us and, consequently, how we expect governments to make policies that support that world view.

> "Governments face the complex challenge of finding the right balance between the competing demands on natural and social resources, without sacrificing economic progress."

First, there is the realisation that economic growth alone is not enough: the economic, social and environmental aspects of any action are *interconnected.* Considering only one of these at a time leads to errors in judgment and "unsustainable" outcomes. Focusing only on profit margins, for example, has historically led to social and environmental damages that cost society in the long run. By the same token, taking care of the environment and providing the services that people need depends at least in part on economic resources. Next, the interconnected, or interdependent, nature of sustainable development also calls for going beyond borders – whether they be geographical or institutional – in order to co-ordinate strategies and make good decisions. Problems are rarely easily contained within predefined jurisdictions such as one government agency or a single neighbourhood, and intelligent solutions require cooperation as part of the decision-making process.

Take genetically modified crops, for example. Making decisions on the production, consumption and development of GMOs requires the participation of agriculture, environment, trade, health and research ministries. It requires that these ministries compare evidence and agree on a position within national government so that they can enact workable policies – policies that have the greatest benefit for the least cost. But the need for co-ordination doesn't stop at the national level. Apart from anything else, seeds from genetically modified plants can cross borders, carried by wind or birds, adding an international dimension to the issue. Differing policies between import and export countries leads to confusion and inefficiency in trade, as processed foods containing just one genetically modified ingredient require special labelling and are even banned by some countries. Finally, thinking about human actions has had to undergo a temporal shift: put simply, we should consider the impact of a given choice beyond the short term. If poorly-managed logging leads to the depletion of a forest in the interest of immediate profit, then the overall result is actually a substantial loss: loss of income over the long term, loss of biodiversity, loss of capacity to absorb carbon dioxide, among other things.

An "honest" approach to timelines is also essential to questions of intergenerational equity: the idea that resources, whether economic, environmental or social, should be utilised and distributed fairly across generations. No single generation should bear an undue burden. This is not only a problem of leaving a clean, healthy planet for future generations, but also concerns pressing problems like meeting the medical, financial and social needs of an ageing population.

Pillars of Sustainable Development

At the core of sustainable development is the need to consider "three pillars" *together*: society, the economy and the environment. No matter the context, the basic idea remains the same – people, habitats and economic systems are inter-related. We may be able to ignore that interdependence for a few years or decades, but history has shown that before long we are reminded of it by some type of alarm or crisis.

The fact of the matter is that we depend on ecosystems and the services they provide in order to do what we do: run businesses, build communities, feed our populations and m uch more. W hether we consider the more obvious, immediately vital examples – the need for soil that can grow food or for clean water to drink – or the less obvious but equally significant things like oxygen production during photosynthesis or waste processing by bacterial decomposers, we cannot avoid the conclusion that we depend on the environment for our existence. If we damage or destroy the capacity of the environment to provide these services we may face consequences for which we are completely unprepared.

In the same way, the long-term stability and success of societies rely on a healthy and productive population. A society (or communities within a larger society) that faces unrest, poverty and disease will not develop in the long term: social well-being and economic well-being feed off each other, and the whole game depends on a healthy biosphere in which to exist.

Understanding the complex connections and interdependence of the three pillars requires some effort, and the effort has to be constant. Whether we're talking about the duration of political cycles or the length of time the media focuses on a particular issue, the question of our collective attention span is an important one for sustainable development.

Trade-offs

With tens of millions of inhabitants concentrated in a limited space, today's mega-cities struggle to balance the needs of the population with the capacity of the existing infrastructures. Juggling the complex web of activities in urban environments is an ideal place to start thinking through the trade-offs that sustainable development can imply. For instance, everybody might agree that traffic is a nightmare, but making changes to improve the situation inevitably affects many people in a variety of ways, not all positive. Should the city discourage car travel but risk overloading public transport? Should it introduce measures to make traffic move more quickly and risk attracting more vehicles onto the roads? Calculating the financial costs of transport policies is relatively straightforward, but predicting the personal choices and behaviours of those using the urban space is much less so. What will city dwellers and commuters actually decide to do? For example, if the bus service improves, will it attract car drivers or people who might otherwise have walked?

The lesson here is not that it's impossible to improve things, but that improvement means thinking through the links among a number of factors. Less traffic equals shorter travel times and easier movement. Better air quality means a healthier population. The trade-offs, such as taxes or tolls, in exchange for overall improvement of the urban space are being tested in London, Singapore and other cities. The debate as to the success or failure of such schemes shows in a concrete fashion what's at stake. The environmental impacts may seem clear, but what about social equity – the rich can afford to pay a congestion charge that poorer people can't – or the economic impact on shops and other businesses?

On a personal level, the choices may not be so clear-cut either. Imagine you want to avoid supporting the use of pesticides so you choose to buy only organic produce. However, the only organic grocery store in your city is too far to walk or cycle to. Fossil fuels have to be burned to get you there and back. Likewise, you may want to support local producers and avoid the damage air transport causes. But fl ying flowers to the UK from Africa for instance may cause less harm than importing flowers from nearby Holland that needed heated greenhouses and intensive fertilizer use. And horticulture may benefit more people in Africa than in the Netherlands. In a perfect world, making good choices would be an easier, more coherent task; in the meantime, the concept of sustainable development is helpful for balancing out the vast number of variables and optimising our decisions.

Process and End Result

So, is sustainable development a kind of guiding principle, as many of its supporters would argue? Or rather a concrete goal or set of goals that can be measured, evaluated and deemed "achieved"? Looking at the massive body of literature on the subject reveals plenty of support for both these points of view and several other possibilities. Really, t hough, there is no o bligation to choose among these o ptions. Whether we are talking a bout the a bolition of slavery, universal education, democracy or any of the "sea changes" that previous generations have undergone, we are always in a constant process of translating big ideas into concrete practices. And this always involves multiple experiments, learning, failures, mistakes and a constant effort at adapting and refining our methods.

Sustainable development is also a means for considering the relationships of things to each other in order to propose viable solutions. As the Brundtland report puts it, "sustainable development is not a fixed state of harmony but rather a process of change…" It is a way of forcing ourselves to look at factors we might rather ignore in favour of short-term benefit, as in the case of a polluting industry that worries primarily about this year's profits, or a pension plan that doesn't account for the increase in the number of retirees relative to the number of subscribers.

Brice Lalonde, former Minister of the Environment in France, offers the following definition: *"To me, it refers to how the economy should enable us to live better lives while improving our environment and our societies, from now on and within a globalised world."* In this view, sustainable development frames the possibilities for progress: the economy is a vehicle that helps us reach the overall, collective, goal of improving quality of life *globally.* Success comes through putting all three pillars on the same progressive trajectory, or path.

It might be useful, then, to see the advent of sustainable development as a significant change in how people and governments perceive their activities, their roles and responsibilities: from primary emphasis on increasing material wealth to a more complex, interconnected model of the human development process.

Sustainable development is therefore:

- a conceptual framework: a way of changing the predominant world view to one that is more holistic and balanced;
- a process: a way of applying the principles of integration - across space and time - to all decisions; and
- % an end goal: identifying and fixing the specific problems of resource depletion, health care, social exclusion, poverty, unemployment, etc.

Society, the environment and the economy - doesn't that cover just about everything? One of the first things we notice when trying to understand sustainable development is the vastness of the topic. Taking into account the economic, social and environmental aspects of development can ultimately include a wide variety of concepts, policies and projects. So wide, some would say, that it loses its usefulness as a concept.

This could in part explain why, in spite of its popularity and rapid acceptance by some members of government, civil society, countless companies and many cities, the concept of sustainable development has not yet translated into widespread changes in either behaviours or policy, and this after more than a decade of efforts. Early supporters of the concept had hoped for rapid progress, but the complexity of the problems at hand, their reach across cities, regions and beyond national borders, and the difficulties inherent in changing people's perceptions and actions have all contributed to frustrating those hopes.

Adding this level of complexity to decision-making processes most probably necessitates changes in previous patterns of behaviour – whether at the level of individual consumption or international law. And change is almost never easy, even when it is obviously necessary. It is particularly difficult when it might involve real or perceived sacrifices on the part of one "pillar", industry, country or generation in favour of another. It is still quite common to hear that sustainable development is primarily about the environment. And while it is true that the concept grew out of thinking about the dangers of environmentally unsustainable practices such as the damage done to ozone layer by CFCs or the damage to soils and water supplies due to pesticides, sustainable development has also always included the social dimension.

In any case, to get caught up in an argument over whether sustainable development is more about the environment or about people is to miss the point: it is the connection of humans, their economies and societies to the ecosystems that support them which defines sustainable development. "Environmental problems are really social problems anyway," said Sir Edmund Hilary, the first man to conquer Mount Everest. "They begin with people as the cause and end with people as the victims."

So really we can see sustainable development as a big theory, a process, or as practical guidelines for making solid development decisions that do not blindly seek growth in one area only to cause damage in another. We can choose to support any or all of these positions, provided that we have the information we need to make honest assessments about our activities and their impact – and make some of the "tough" decisions that good management often requires. Applying the principles of sustainable development is really nothing m ore than applying the principles of sound management to all our resources, like w e w ould if we wanted to create a prosperous business or build a new house. Rather than overlook potential conflicts, we can plan ahead, integrating considerations of what counts from the beginning. Of course this is easier said than done:

Spending money now to prevent something that "might" happen in the future is a challenge for us. Just as spending money to fix a bad situation "elsewhere" is also tough. Really, though, the future is right around the corner, and in our globalised world what seems far can become very suddenly close. By following the example of the ever-increasing number of individuals, businesses and governments who make planning decisions within a sustainable development framework, we ensure ourselves and our children a brighter future.

Challenges of a Global World

Life in Ahoto in the s tate of Jigawa, N igeria has followed the same rhythm for centuries. In this village of thatched mud huts, farmers eke out a subsistence income from the difficult lands to the south of the Sahara desert. But things have begun to change recently: solar power has come to Ahoto and brought with it substantial improvements people's lives.

Garba Bello, head of the village, is very happy about these changes. As a recipient of one of the household lighting systems (costing a bout four dollars per month), he enjoys what solar lighting has done for his household, and most of all for his village. “The difference is great,” he says. “People now go out at night and chat. Before, you could not even see your neighbour’s house in the night.”

The solar power project has brought more than just light to Ahoto and the two other villages participating in the region. A new shopping area is driving business development and creating much needed economic activity. Educational opportunities are also on the rise: women now attend classes at night and children can do their homework. A collaborative effort by NGOs, the state government and foreign aid, the Jigawa State project takes a promising use of alternative energy a step further than previous projects that focused on only one use, such as pumping water. By attempting to address *all* of a village’s energy needs, from education and commerce t o security and women’s advancement – the project gives participants the means to move forward simultaneously in all areas of their development.

The benefits extend beyond the social and economic to questions of health. Villagers now have access to clean water from more efficient solar pumps that draw water from deeper non-polluted sources and distribute it to household and communal taps. Easier access to relatively l ow-cost potable water also frees u p considerable amounts of time that used to go to collecting water by bucket or hand pumping from wells. Activities that had to end at sunset can now continue, and in healthier circumstances. Dangerous and dirty kerosene lamps are now rarely lit. This has significant implications for health. About 1.5 million people die prematurely each year from the effects of indoor pollution from burning wood, charcoal and waste, more than from malaria, almost as many as from tuberculosis and almost half as many as from HIV/AIDS.

It is a solution that is elegant in its simplicity - bypassing, or “leapfrogging”, traditional technology to move directly to one that is cleaner and far more sustainable. Yet projects like Ahoto’s are still too few compared to the vast energy needs of the developing world, where, on present trends 1.4 billion will still be without access to mains electricity in 2030.

Old Process on a New Scale

The phenomenon of globalisation has received a considerable amount of attention in recent years as social scientists, political pundits and cultural critics have tried to explain how it has transformed our world. Is it really new, though? For as long as explorers have had the means to cover great distances, people have sought to know, understand and profit from what lay beyond the familiarity of their own communities.

The great periods of exploration and colonisation attest to this desire, one that combines many different motivations. To learn about the world, to seek out better means of survival when local methods had failed, to seek fame and fortune, to trade for what one lacked, to bring glory to the state - these various driving forces interacted and pushed human societies forward to a world that became more and more connected over time. Nowadays, globalisation is not about a few rich countries trading with far-off lands. Geopolitics, technology and finance have transformed consumption and production patterns across the globe. Over the past decade alone, around a billion workers have joined the global marketplace. Better communication tools and falling transport costs have expanded the range of goods and services in national markets. The combination of a greater supply of cheap labour alongside technologies that facilitate trade means that "value chains" – the numerous steps involved in transforming materials, knowledge and labour into saleable products – are spread across the globe. What is new about the globalisation of the last 30 years or so is that we are nearing a point where connection is not the *exception* but the *rule*.

According to the World Trade Organization, international trade has consistently increased with yearly growth rates of around 6% over the last decade. China is a leader here, with merchandising export increases of 27%, while developing countries' share of world merchandise exports reached an all time record of 36%. It is a fact that our economies are now dependent on international exchanges for their continued progress.

Whether we are talking about trade and investment, politics or culture, examples of linkages and interdependence surround us. We need only think of the food on our plates, the clothes on our backs or a website like YouTube: the sources of what fills our daily lives are multiple and geographically diverse. We sample the world's offerings every day, rarely realising how all of those connections have come to be, or how they interact. These changes in our daily lives are related to international developments with a substantially increased movement of money and things: trade rules have been "liberalised" or modified to encourage international competition; corporations have expanded beyond their country of origin to new markets around the globe. All this moving and mixing has opened up possibilities for exchange, commercial expansion and overall growth, making the world, on average, a richer place.Recent years have indeed witnessed striking changes in the global economic landscape, confirming the role of trade as a driving force in economic development and providing an indication of the potential for further trade liberalisation, under the right conditions, to benefit the global economy broadly. This "new" global dimension – economic, political and social – offers seemingly endless opportunities. But these opportunities are not equally available to all, and means to restore the balance have to be found. Nobel Prize economist Joseph Stiglitz has written recently about one of these, the fact that economic globalisation has left politics trying to play catch up. He points out that globalisation has in some ways changed the role of the nation-state since so many important issues now reach beyond the country level. In spite of this shift, Stiglitz notes, "there has yet to be created at the international level the kinds of democratic global institutions that can deal effectively with the problems globalization has created."

A Two-tier World

Globalisation has increased our ties across geographical boundaries and perhaps transformed the way we think of «the world». Yet for all our increased connections, we obviously don't share the same circumstances, lifestyles or opportunities. As long as we have lived in large communities, the "haves" and the "have nots" have existed side by side. In today's media-rich culture, it is hard not to be aware of the glaring disparities in living standards in different parts of the world, even as we all participate in the same global economy.

Where we are born, grow up and live makes a big difference. An average child growing up today in Europe has vaccinations, dental care and educational opportunities, not to mention a more than adequate diet. He or she can look forward to higher education, travel, employment and retirement provided, at least in part, by a stable government. Economic growth in such OECD countries hovers around 2.5%, enough to maintain and hopefully continue improving quality of life provided that resources are managed wisely.

Some other countries have, after long periods of no or slow growth, recently gained speed – at least in terms of increases in their Gross Domestic Product (GDP). Still, many of their citizens are living "in another world". In India, where growth has averaged 8.5% for the last 4 years, 300 million people still live on less than the equivalent of a dollar per day. Africa has shown encouraging growth too – averaging above 5% per year for the continent as a whole in 2007, the fourth successive year of record growth – yet life expectancy in many African countries remains shockingly l ow. In Swaziland, people live on average 39.6 years – less than half the average lifespan in Japan, the country with the highest average. Indeed, the different speeds at which countries are meeting basic development goals leaves us asking: Will economic growth translate into equal opportunities for citizens in the near future? What else can be done to make sure that more of the world's underdeveloped communities gain ground in as fast and efficient a way as possible?

National Growth has Global Consequences

For China, India and the other emerging economies, growth is happening rapidly, bringing with it both the positive and negative consequences of intensified production and increased economic activity. Because of their sheer size, the choices these countries make as to how to direct their growth have a huge impact worldwide. Media everywhere noted with gravity a symbolic passage in 2007: China is now the world's largest producer of CO_2 emissions. But w e cannot forget that its emissions per capita are still a fraction of those of OECD countries. The principle of "shared but differentiated responsibility" between developed and developing countries tries to take this into account. Outlined at the 2002 Johannesburg Summit on Sustainable Development, the principle recognises historical differences in the contributions of developed and developing states to global environmental problems, and differences in their respective economic and technical capacity to tackle these problems.

Environmental consequences such as climate change respect no borders, driving home the necessity for a global perspective on pollution. To reach their current sizes of about $14 trillion and $16 trillion annual GDP respectively, the United States and European economies drew heavily on natural resources and depended almost entirely on fossil fuels. Today's changing climate is largely due to the rich countries' historical emissions. The economic model that drove this development had serious consequences for the environment, such as the permanent destruction of species and ecosystems and an increase in atmospheric carbon dioxide that m ost scientists believe is already changing our climate. Now the model, and its consequences, are being replicated by other countries at an accelerated pace.

The global nature of our economy means that we are increasingly linked to other countries. Disruptions half-way across the world come home in a dramatic way when they drive up the price of food or petrol at home, or result in a deployment of armed forces. People living in regions where economic prospects are dim may choose, by whatever means possible, to emigrate to wealthier destinations. While the positive impacts of immigration in OECD countries are well-established – providing much-needed labour for example – humanitarian and economic emigration can put a burden on countries' social systems of both the country people leave and the one they go to, especially in situations of crisis and when the "host" country is a developing country.

By 2030, the world population is expected to reach 8.2 billion people from 6.5 billion today. Such projections from the most recent OECD Environmental Outlook can seem like daunting increases when we consider that the world's resources are already stretched to capacity in many respects. Where will that growth take place? A large part of it will be in the fast-developing economies of Brazil, Russia, India, Indonesia, China and South Africa – known as the BRIICS. What form will that growth take? How can we all shift to more sustainable models of growth?

A Level Playing Field

Those who criticise the environmental record of rapidly developing countries run into an interesting debate regarding the "right" of developing countries to pollute or to have access to more advanced, less damaging technologies. Europe and the United States spent several hundred years practicing rampant deforestation and industrial pollution before putting in place tight regulation. Why should China and Indonesia have to play the game by different rules from those followed by developed countries in the past?

There can be no moral grounds for expecting China and India selectively to curb their economic growth simply because world energy demand is rising unacceptably, with associated risks of supply interruptions, high prices and damage to the environment. These are global problems to be tackled on a global basis. Indeed it is often perceived as unfair for rich countries to lecture poor countries on resource use when the developed world is responsible, by its size, history and volume of activity, for the majority of resource consumption and the problems that ensue from irresponsible development. While all major emitters must act, the developed countries need to take the lead in addressing climate change. With global issues such as ozone depletion, climate change and biodiversity loss, everyone feels the effects of development when it is unsustainable; everyone should also feel the benefits when it is sustainable. Whether we are talking about people's quality of life or sound natural resource management, success depends on the participation of countries, regions and localities at all stages of development.

So then the question becomes one of *how* to fairly share the burdens of achieving well-managed growth. Developing countries have to cope with climate change and other problems they did not create, and do not have the same means as developed countries in tackling them. The developed countries can help by providing technologies, finance and know-how to tackle these issues, over and above regular development assistance. The international community has engaged in various forms of development aid for over half a century; billions of dollars have been spent on different types of projects designed to spur growth and improve living standards in the poorer countries. The current international consensus is that each OECD country spends 0.7% of gross national income (GNI) on foreign aid versus 0.3% at present in order to reach global development targets such as the Millennium Development Goals. Aid to Africa alone is expected to reach $51 billion by 2010 from $40 billion in 2006. But making sure that aid is directed to sustainable projects adds another layer of complication.

Closing the Development Gap Sustainably

Meeting the needs of today without diminishing the capacity of future generations to meet their needs: sometimes discussions on sustainable development have tended to focus more on the second half of this of this phrase – the effect of our actions on the future – than on the first half. Yet meeting the needs of today is anything but obvious, easy or conflict-free. wealthier and poorer countries – must figure among its most pressing projects.

"It is not easy for men to rise whose qualities are thwarted by poverty," observed the 1st century Roman poet Juvenal. The question of what causes poverty and what can reduce or eradicate it has long been a source of much debate, one of the fundamental questions facing the human community. We all have some idea or image in our heads of what constitutes poverty. It's not just a question of possessions – poor people in rich countries own more things than most people elsewhere. Research on the topic points to a more complex combination of the material, social and political aspects of poverty, where lack of access to information, political participation, health care and education among other things contributes to blocking the dynamic that would make lasting development possible. Being ill, hungry, or having to flee violence forces people to redirect their energy to the act of survival, without the luxury of long-term considerations. Addressing the basic needs of the world's poorest people would go a long way in fostering development in today's global economy, but it would obviously require a global approach.

Early promoters of sustainable development realised that to make the substantial changes required to produce meaningful results would require a global effort. No person, municipality, region or even country could alone transform the ideas and practices driving development. The increasingly important role of international agreements on the common concerns of the global community – trade, multinationals, poverty reduction to name only a few – attest to the need for international arena to solve problems of global significance.

The UN, OECD and other international organisations are struggling to bring sustainable development to the f orefront.Other organisations including the World Bank, Intèrnational Monetary Fund and World Tr a de Organisation are seeking means to incorporate sustainability as a basic principle in their economic operations. These institutions have brought national governments with a diverse range of views and means together to the same table to hash out their differences in the interest of improving development practices. At the same time, local and regional governments are joining forces to compare their experiences and work together, often across great geographical distance.

All in all, governments are starting to realise that they need to undertake a more collaborative, open approach to problems that are more crosscutting. The idea of being able to improve the lives of the poorest through global action has taken hold over the course of this century, culminating with the Millennium Development Goals -an effort to attack the problem in a co-ordinated way and on a global scale. Originating in the OECD guidelines for development, the MDGs, as they are known, represent a concerted effort on the part of the world community to address the persistent problems of underdevelopment.

The Millennium Development Goals

Officially established in 2000 at the UN Millennium Summit, the Millennium Development Goals identify 8 development goals and under these 18 concrete targets to be reached by 2015.

Agreed by 192 UN Member states, they represent a global agreement to achieve results in the most critical areas of human progress.

1. Eradicate extreme poverty and hunger

Reduce by half the proportion of people living on less than a dollar a day.

Reduce by half the proportion of people who suffer from hunger.

2. Achieve universal primary education

Ensure that all boys and girls complete a full course of primary schooling.

3. Promote gender equality and empower women

Eliminate gender disparity in primary and secondary education preferably by 2005, and in all levels by 2015.

4. Reduce child mortality

Reduce by two-thirds the mortality rate among children under five.

5. Improve maternal health

Reduce, by three-quarters the maternal mortality ratio.

6. Combat HIV/AIDS, malaria and other diseases

Halt and begin to reverse the spread of HIV/AIDS.

Halt and begin to reverse the incidence of malaria and other major diseases.

7. Ensure environmental sustainability

Integrate the principles of sustainable development into country policies and programmes; reverse loss of environmental resources.

Reduce by half the proportion of people without sustainable access to safe drinking water.

Achieve significant improvement in lives of at least 100 million slum dwellers, by 2020.

8. Develop a global partnership for development

Develop further an open, rule-based, predictable, non-discriminatory trading and financial system.

Address the least developed countries' special needs.

Address the special needs of landlocked countries and small island developing states.

Deal comprehensively with developing countries' debt problems.

In cooperation with the developing countries, develop decent and productive work for youth.

In cooperation with pharmaceutical companies, provide access to affordable essential drugs in developing countries.

In cooperation with the private sector, make available the benefits of new technologies - especially information and communications technologies.

The MDG Monitor tracks the progress towards achieving these goals. It gives an overview of the principle targets that come under each goal, indicators for measuring progress and examples of success stories.

Tools for Sustainable Growth

With the exception of a few rapidly growing economies, growth in the developing countries has on the whole been inconsistent and insufficient to bridge the huge differences in living standards within these countries and compared to the developed countries. Recent indicators show that Sub-Saharan Africa has begun to see the rates of growth comparable to the rest of the world, albeit from a low starting point, but this has not yet translated into great gains in a number of critical areas. For example, the number of people in Sub-Saharan Africa with access to drinking water increased by 10 million per year over 1990-2004. However, population sizes have grown even faster, so the number of people without access has increased by about 60 million.

Each country's historical, economic, social and political context is unique, but the basic principles of sustainable development apply to all. Economic growth is essential, but growth alone, without understanding all the factors that contribute to well-being including social, environmental, institutional and cultural considerations, does not produce sustainable poverty reduction. While it is true that economic growth generally correlates with overall improvements in quality of life, higher levels of education and life expectancy at the country level, this does not tell us:

- how this growth is achieved
- whether or not it is lasting

- who benefits and who might be left behind.

Countries with highly valued natural resources such as diamonds, metals or oil have the means to increase overall economic development by selling these resources on world markets. Still, this may result in no improvement in people's lives if this income remains in the hands of very few and is not used in any way that benefits the population. If these resources are non-renewable, or poorly-managed, then the income they produce will at some point cease to be a source of growth – unless profits from them are reinvested in other projects or funds that are sustainable over time. Finally, activities that are profitable today may be degrading the environment for tomorrow. In sum, short-term growth may mean nothing in terms of long-term stability, and it can produce net environmental and social loss if the stocks and the capital generated from them are not managed *sustainably*.

Pro-poor Growth

So the question is how to create growth that allows the poor to achieve real, lasting advances. Economists and development theorists call this *pro-poor growth*. According to this way of thinking, it is not enough to achieve average growth rates of a certain percentage. Growth should specifically benefit poor women and men and allow them to reap the benefits of increases in economic activity and income so that they can access a path of consistent improvement in their living conditions.

What exactly are the means of meeting development goals? We all know some of the basic elements such as capital, health, education, technology. Approaches to providing aid and fostering growth are diverse. Development specialists describe three major avenues: official development assistance or ODA, foreign direct investment (FDI), and trade. Though these are distinct categories in a definitional sense, they work together in practical terms. For example, ODA funds might be directed to measures intended to attract FDI or develop trade, as in the case of Aid for Trade. New forms of aid have also begun to play an important part in helping achieve development targets. Large private foundations such as the Bill and Melinda Gates foundation and public-private partnerships like the Global Fund to fight AIDS, Malaria and Tuberculosis and the Global Alliance for Vaccines and Immunization (GAVI) have had a significant impact on how aid programmes are conceptualised and implemented, even though they represent a relatively small percentage of the funding pie.

Donor countries and agencies have certainly become aware of the importance of sustainability issues and are working to ensure that environmental considerations are integrated into the Poverty Reduction Strategies of recipient countries. The United Nations Environment Programme (UNEP) and the UN Development Programme (UNDP) oversee many of these efforts as part of the Poverty-Environment Initiative. In turn, the OECD's Development Assistance Committee (DAC) keeps track of these projects, in accordance with the Paris Declaration on Aid Effectiveness. The objective is to ensure that environmental concerns are integrated into development strategies, although for the time being progress is uneven.

Another example of the international effort to ensure that environmental considerations are included in development initiatives is Strategic Environmental Assessment (SEA). While Environmental Impact Statements have long been required by donor countries, they often represent a last item on a checklist for a project, an approach that leads to conflict of interest and missed opportunities. Since 2001, both donor and recipient countries have been passing legislation to undertake SEA for development programmes likely to have an impact on the environment. With SEAs, environmental considerations are integrated from the conception of national or regional Poverty Reduction Strategies. But we still have a long way to go before sustainability assessments of all three pillars are routinely conducted and Poverty Reduction Strategies are transformed into Sustainable Development Strategies.

Making Aid Count

There is a need to ensure that the development objectives of countries giving aid and those receiving it are co-ordinated and mutually reinforcing. In March 2005, representatives from nongovernmental agencies and over 100 countries – both donors and recipients of aid – came together to sign an international a greement in this direction: The Paris Declaration on Aid Effectiveness.

Co-ordinating the efforts of different donor and recipient governments is challenge enough – but add to that all the other actors involved in the development process such as NGOs, the media, or financial institutions, and we soon see why the infusion of capital is simply not sufficient. Donors must be organised and coherent in their approach – what is called "harmonisation." They should base their efforts on the needs expressed by recipient governments' national strategies. Donors, recipients and professionals working in project implementation realise that without better co-ordination, local engagement and accountability, aid is likely to fall short of its targets. The Paris Declaration on Aid Effectiveness of 2005, an agreement signed by donors and recipient governments as well as multilateral aid organisations, reflects this commitment to a more coherent, realistic approach to meeting development goals.

Realising and articulating the need for better, more focused coordination is an important step, but it is only a beginning. Closing the development gap depends on building sustainable, healthy societies – removing the barriers that impede progress and bringing into the discussion such things as human rights, gender equality, peace and security. These more complex and diffuse elements pose a challenge for monitoring, but the need for policy coherence on these items is included in the reviews (know as "peer reviews") the OECD carries out of its member countries' development assistance efforts.

Going Forward

In spite of the amount of aid, investment and trade-related growth, the development gap remains. One reason for this is a lack of co-ordination. Sometimes policy aims conflict, as when donor countries provide aid for health systems and at the same time try to attract doctors and nurses from the developing world.

Countries are starting to address what has to happen to make aid, trade, investment and other economic policies perform better in order to a chieve lasting development r esults. In specialist language, this is called "policy coherence for development" which means making sure that the economic objectives of donor countries are coherent and do not undermine each other. For example, subsidies to domestic farmers or fish ers do not negate gains in o pening world markets, export credits or investment incentives do not conflict with the goals of development assistance policies do not interfere with building human and social capital, and so on.

According to a much-quoted saying, "Give a man a fish and you feed him for a day. Teach him how to fish and you feed him for a lifetime." But is this true? What if he overfish es? Or more efficient boats from elsewhere take all the accessible stocks? Or pesticides are washed into the breeding grounds and drive the fish away? Teaching "how to fish " involves much more than knowing how to cast a net. We have to understand the critical nature of linkages – how things relate to each other. And it is here that applying the principles of sustainability throughout the development process takes on its full meaning. The goal is not for the developing world to "catch up" with the bad habits of the industrialised countries, but rather f or the developing and developed countries to co-operate in instituting sustainable growth across the board. If we are to put both richer and poorer countries on a path to development that lasts, we all have to start fish ing sustainably.

The Future is Now

Straddling the border of Poland and Belarus is a magical place, seemingly untouched for thousands of years. In the spring, wildflowers bloom under majestic oaks, and animals give birth to their young in the last remaining fragment of a primeval forest that once covered almost all of western Europe. The preservation of this particular region began centuries ago, when tsars and princes reserved the land as a private hunting ground for the elusive and increasingly rare wisent, the European bison. During the First World War, the forest and its inhabitants were again in serious danger: logging mills were built, and the last wild wisent was killed by a poacher in 1919. It seemed this last corner would go the way of wild areas on the rest of the European continent, with the virgin forest and its large mammals lost forever.

As soon as peace returned, however, determined conservationists went to work, and in 1932 the Bialowieza National Park was established. In the decades since, this unique ecosystem has been recognised as a UNESCO World Heritage Site and Biosphere Reserve. The wisent were reintroduced in 1952 from the small population surviving in zoos. Today, the bison population is healthy at about 250 individuals, in addition to other large mammals such as elk, deer, wolves, wild horses and over 100 species of birds. Each year, 100 000 tourists visit the small area of the forest open to the public, for a glimpse at this rare and wonderful ecosystem. At the end of 19th century, the primeval forests of western Europe had been gone for generations, and in the United States the last borders of virgin forests were being logged. Species like the American bison had been hunted down to the last few hundred; others like aurochs and the great auk were gone forever. But a movement to protect the last wild places swept across Europe and the Americas, and over the 20th century thousands of square kilometres were set aside, protected in one way or another, for future generations.

By establishing national parks we narrowly escaped the permanent loss of many species and ecosystems. Now we turn to our future and wonder – what is it that we need to protect, or risk losing forever? With economic development and urbanisation racing ahead, how do we ensure that we're giving future generations a fair chance for the kind of lifestyle we've enjoyed? As the pace of human activity and impact increases, today's adults might even worry for the stability of our own future.

A century ago it seemed enough to set aside areas of special habitat. Today, we know that not only have we used up certain resources, accumulated national debt and released long-lasting pollution into water, air and soil, we're even changing the climate on which our lives depend. Clearly, it is time for another type of conservation movement: one that helps us manage what is important to our well-being and that of future generations, responsibly and sustainably.

A Fair Share between Generations

When the concept of sustainable development was fifirst articulated in the Brundtland Report, fairness to future generations was a central tenet. This concept is sometimes called *intergenerational equity.* While relations between nations are regulated with laws and agreements, people who will live in the future can't defend their rights, even though their well-being is affected by our actions. We therefore have a duty to protect their interests, even at the cost of potential short-term gains to us.

The problem is not only one for generations at some far off time disconnected from the present: in reality, the future is as early as the child born five minutes from now. Managing systems for the long term is not an altruistic notion. Our interest is in the future because we are going to spend the rest of our lives there, to paraphrase American inventor Charles Kettering.

This is obviously a huge challenge, involving choices that we as citizens have to inform ourselves about. Take some of the most hotly debated topics such as health, pensions or public debt. You often hear that health expenditures will rise because of population ageing – the "grey dependency ratio" shown in the graph. But analyses carried out by the OECD present a more complicated picture. Although health costs rise with age, the average cost per individual in older age groups should fall over time in part because people are not only living longer, they're staying healthier longer. And they'll be getting pensions longer, too. Should it be up to individuals to make sure they have enough to live on at retirement, or should we tackle this issue as a community? Or what about public debt? Is it merely a burden we're passing on to our children or are the infrastructures, education or other services it pays for an investment in their future?

And what about our stewardship of the earth's land and resources? Outside of those lands set aside for protection, we have a history of exploiting resources through intensive activity. Can we manage most or all our forests, wetlands and oceans so that they continue to provide the riches we rely on? Are the habitat changes caused by our development endangering species that our descendents might value for aesthetic and philosophical reasons, or even practical uses like medicine and agriculture?

It is as if suddenly, after tens of thousands of years of human progress and all the activity that goes with it, we have finally grown into our planet: we can reach its most remote corner; we circle it in a day; we can re-direct its rivers and climb its highest peaks. It also seems that we could soon grow out of it if we do not proceed carefully. Indeed, the planet is showing signs of reaching critical thresholds in all of its major systems. Climate change, species loss and pollution are evidence that the world's capacity to handle what humans generate is close to full. Lest this sound like a purely "environmental" problem, keep the following in mind: the changes that occur as a result of higher temperatures, natural disasters or losing an important insect in the food chain all have profound effects on individual human beings and society as a whole. If there is one realisation that seems to have taken hold since the advent of sustainable development, it is that the environment, the people who inhabit it, and the economies and cultures they thrive on depend on each other.

> "The remaining environmental challenges are of an increasingly complex or global nature, and their impacts may only become apparent over long timeframes. Among the most urgent of these challenges for both OECD and non-OECD countries are climate change, biodiversity loss, the unsustainable management of water resources and the health impacts of pollution and hazardous chemicals. We are not managing our environment in a sustainable manner."

Learning to be Sustainable

For over a hundred years, forestry schools have been teaching how to manage forest r esources. Knowing how fast a certain species of tree will grow in a given climate, it is possible to calculate the sustainable yield. In a tree plantation, the trees can even be treated as an 80-year agricultural crop with the same species covering thousands of hectares, and they can be harvested indefinitely if the soil is fertilised and climate conditions don't change.

This type of forest, with trees of one age and one species, is not a useful habitat for very many other species, but even tree plantations can be managed for maximum biodiversity.

A bigger challenge today is managing fish eries resources: the main way we have of tracking these populations is based on catch. How can we know whether we're overharvesting the resource? Improvements in equipment and methods engineered over the last half century allow for greater catch – giving the impression at a given point in time that the population is healthier than ever – until it crashes. This is precisely what has happened in some of the world's best fish eries, such as the Grand Banks off the coast of Newfoundland in Canada.

At least one-quarter of marine fish stocks are overharvested. The quantity offish caught increased until the 1980s but is now declining because of the shortage of stocks. In many sea areas, the total weight offish available to be captured is less than a tenth of that available before the onset of industrial fish ing. Inland fish eries, especially important for providing high-quality diets for the poor, have also declined due to overfish ing, changes to habitats and withdrawal of fresh water. What can be done to conserve these valuable aquatic resources for future generations? Marine biologists, fish ermen and policy experts have proposed several possible solutions to make sure that fish stocks w ill be available w ell into the f uture: quotas are imposed for each species, in hopes that enough are left in the waters to reproduce; Marine Protection Areas are established, with strictly no fish ing, as a base from which populations can grow. Naturally, these policies are only effective when they are fully enforced. For the fish ers themselves, government programmes buy back boats, offer professional reconversion programmes and try generally to support communities where there are simply too many fish ers for the resource.

Finally, we can replace wild fish with a more easily-managed resource: fish from aquaculture. Aquaculture currently provides almost 40% of the fish and shellfish we eat, but it has its limits as well. Raising so many fish in such small quarters makes the risk of infection so high that antibiotics have to be used. Escaped fish interbreed with wild populations, endangering their genetic diversity, and pollution from the fish food and wastes flows easily into surrounding waters. All of these problems will require continued attention and investment in order for aquaculture to become a truly sustainable solution.

With hindsight, the tragedy of Grand Banks seems avoidable. Yet other fish eries are likely in the same situation today as Grand Banks was in 1991, with a catch that seems stable or declining only slowly, but is really nearing or already past its critical threshold. Globally, we still have a very hard time following Larry Tremblett's advice. Even if we know that a crisis may be just around the corner, we all find it difficult to step back and make major changes.

Now is the Time to Act

One major barrier to making changes is the degree of uncertainty that comes with planning for the future: there is no experiment to "prove" what the exact consequences of unsustainable choices will be. In the case of climate change, we don't know for certain what amount of CO_2 in our atmosphere will trigger serious and possibly irreversible damage – our Earth is our only laboratory on this one. The same is true for biodiversity loss or any of the other resource-management issues that we face. This uncertainty is sometimes taken as an excuse not to make investments in better, cleaner sustainable practices: why pay all that money if we're not *sure* we have to? And yet that kind of reasoning leaves us vulnerable to a big surprise once there is overwhelming evidence that a systems failure of some type will be devastating, it is likely to be too late to muster the means to avoid it. Sure, some miracle solution might come along, but do we really want to take that risk? What sustainable development argues for is to mitigate those risks now, in ways that enhance our present, as well as preserve our future.

When faced with the prospect of major changes to our environment, a common response is to say "human beings are so resourceful; we'll find a way to deal with that problem when the time comes." Indeed, new methods and technologies can lead to decreased reliance on natural resources, allowing us to give the planet a break, but they can also increase pressure on natural ecosystems or create new worries of their own, as in the fish eries example. Those technological wonders that we hope will appear in time to save us take years to develop, and there is really no break between a "now" in which we can procrastinate and a "future" when we can start thinking about solutions. Waiting until a problem presents serious consequences is clearly not the best way of managing things.

Even if we can avoid the most drastic outcomes, emergency solutions tend to be very expensive. And often we can only mitigate the negative effects of a problem, rather than erase them. Whether the crisis is starvation, pollution or flooding, those who are affected at the moment of crisis suffer the consequences of a lack of long-term planning. And in the case of species loss, there is no possible solution what is gone is gone forever.

Planning for the Future

We want to guarantee that our actions today won't leave behind unsolvable problems and a planet whose capacity to meet the needs of its inhabitants has been depleted. And yet we don't have a crystal ball to look into to see who the people of the future are, how many they are and what resources they require to lead fulfilled lives.

What we *can* do is project into the future using computer models that build on today's situation to try to predict future conditions. Such models can be used to forecast the availability of various social and natural resources, from health care and pensions to fossil fuels and fish. They also forecast the demand for these resources, based on several factors: population growth, economic growth and technology choices. They can give us essential information on what could happen if we do not make the necessary changes. Imagine possible scenarios for the year 2050: in one, the human population has grown to 9 billion and our societies have continued with fossil fuel-intensive development. As natural gas resources have dwindled coal's share in electricity generation has grown. Increased production has led to increased electricity demand, and more people are driving cars. As a result, annual global greenhouse gas emissions have increased by over 50%, going nearly 47 Gigatonnes – that is, billion tonnes – in 2005 to over 70 Gigatonnes in 2050. The concentration of CO_2 in the atmosphere is above 500 ppm and is still rising.

Another possibility: the same population growth has occurred, but economies have shifted from materials-intensive production to service and information activities. Government policies to mitigate climate change, such as taxing greenhouse gas emissions, have been in place for 40 years. Clean and efficient technologies for generating energy and managing emissions have been rapidly developed and s hared worldwide, and non-fossil sources of energy have a far bigger share in the energy mix. Global greenhouse gas emissions peaked around 2 015 and the atmospheric concentration of CO_2 is just stabilising at 450 ppm.

These are the kinds of scenarios that policy makers are considering as they try to balance today's needs with tomorrow's: what the world will look like if we change little or nothing, and what we can achieve if we undertake concerted, co-ordinated actions. Mathematical equations that take into account population, economic growth and energy consumption are used to project future greenhouse gas emissions. This data is then plugged into an even more complex climate model, revealing, with the best current knowledge we have, the impacts. If we're heading towards the first scenario, we could expect a temperature increase of 4-6°C or more in the long term. In the second scenario, the models indicate a more moderate 2-3°C in the long term. Remember that for humans, a heat wave that is just a few degrees warmer than usual can cause thousands more deaths, as Europe experienced in the summer of 2003. Not to mention the more complex effects that warmer temperatures are already having on glaciers and ice caps and sea level.

Such models do not tell us what as yet unforeseen solutions might appear on the horizon, but they can help us understand the possible consequences of decisions we make now. And these days they are sending a clear message: our current path of development is hurtling us towards major changes, changes that will affect almost every aspect of our lives.

Climate Change and Our Future

Our species, *Homo sapiens*, has established agriculture, cities, writing and an impressive array of technology during the relatively stable climate of the last 10 000 years, since the end of the last ice age. Now evidence shows that we are changing the very climate we depend on, largely because of our dependence on the fossil fuels (first coal, then oil and natural gas) that made the industrial revolution possible. Energy needs will increase in the foreseeable future, as developed countries continue their economic growth and developing countries race to catch up. If governments around the world stick with current policies, the world's energy needs will be well over 50% higher in 2030 than today, with China and India together accounting for almost half the increase in demand.

We are already paying for historic emissions from developed countries with more frequent heat waves and stronger hurricanes. At the current rate, the Arctic waters will be completely ice free in summer by the middle of this century, possibly within ten years. Seas will continue to rise as the warmer water expands and is joined by melt-water from glaciers and ice caps.

For the last two decades the debate about the seriousness of this threat has raged mounting evidence of substantial alteration to the climate on one side and scepticism on the other, with some people dismissing human-caused climate change altogether. Yet the latest scientific evidence overwhelmingly supports the hypothesis of a climate that is already undergoing change due to human activity. All of these changes have potentially huge financial and social costs that make inaction seem illogical, short-sighted and even immoral. For example, the Intergovernmental Panel on Climate Change (IPCC) warns that agricultural production in many African countries and regions could be severely compromised by climate variability and change. The area suitable for agriculture, the length of growing seasons and yield potential, particularly along the margins of semi-arid and arid areas, are expected to decrease. This would further adversely affect food security and exacerbate malnutrition in the continent. In some countries, yields from rain-fed agriculture could be reduced by up to 50% by 2020.

On the other hand, recent projections show that costs of reducing carbon emissions will have a minimal effect on global growth. World GDP is projected to double by 2030 and triple by 2050. Stabilising greenhouse gases in the atmosphere at about 450 ppm CO_2 is by all calculations affordable compared to expected economic growth and to the estimates of cost of inaction. The OECD estinated that this stabilisation would cost a small fraction of accumulated wealth worldwide in the comming decades, possibly less than one tenth of a percent of world GDP growth. It is not cheap, but manageable.

When put in those terms, paying the price to reduce carbon emissions now sounds like a smart choice. Also, the more cooperation there is on a global scale, the lower the costs will be.

Calculating the Costs of Inaction

The cost of making changes has often been cited as the reason why we have not managed to take more comprehensive action to eliminate bad habits. One difficulty arises in trying to calculate and compare these costs. We are quite used to calculating the cost of something new. Say a factory is contemplating changing to a cleaner production process, adding a filter that will reduce emissions of nitrogen oxides (NO_x): first there is the cost of the new equipment itself. To this must be added the costs of stopping production while the modifications are made and the cost of disposing of the old materials.

Determining the cost of inaction, however, requires bringing together a number of previously separate issues, some of might not be easy to put a price tag on, such as health and quality of life. Particulate matter produced by fires, diesel engines and incinerators, among other sources, is known to cause heart and lung disease, cancer and respiratory ailments: 960 000 premature deaths and 9.6 million "years of life lost" worldwide was the estimated figure for the year 2000. Photochemical smog, a result of several emissions present in dense urban areas (NO_x, CO_2, SO_x, and ground level ozone, O_3) also causes respiratory illness, cardiovascular problems and increased mortality.

So how much is this extra pollution costing society? Missed work days for adults and increased asthma treatment for children both cost money to the local and wider economies. Smog also affects the value of real estate, and the growth of plants. These are complex calculations at the local level. At a national level it is estimated that damages from air pollution in the US range between $71 and $277 billion per year. Reducing our emissions of greenhouse gases will cost even more than retrofitting factories to control local air pollutants like NO_x. But the potential costs of inaction on climate change are higher, too.

We will act with resolve and urgency now to meet our shared and multiple objectives of reducing greenhouse gas emissions, improving the global environment, enhancing energy security and cutting air pollution in conjunction with our vigorous efforts to reduce poverty. In response to the political will expressed by leaders of industrialised nations at the G8 Summit at Gleneagles in 2005, the International Energy A gency (IEA) has published a series of scenarios and strategies aimed at meeting environmental goals. The so-called ACT scenarios show that, with the right decisions taken early enough, it is possible to move the energy system onto a more sustainable basis over the next half century, using technologies that are available today or that could become commercially available in the next decade or two. The ACT scenarios only stabilise emissions at 2005 levels.

But returning emissions to 2005 levels may not be enough. The IPCC has concluded that emissions must be reduced by 50% to 85% by 2050 if global warming is to be confined to between 2°C and 2.4°C. The BLUE scenarios look at how this could be done, including the use of technologies that still have to be developed, such as hydrogen fuel cell vehicles. Here again, our choices are many: will we combine the forces of governments, the business community and our own personal choices to make the necessary changes? What the scenarios show is that we *are* capable of reducing our emissions, shifting away from activities that affect the climate and still grow our economies *if* we make a concerted international effort to manage the climate change question sustainably.

Educating Ourselves for a Sustainable Future

Finally, we have to take a close look at what got us into our current situation and think seriously about how to change our habits and methods in a way that will last. Now that we know so much more about the relationship between development and the well-being of people and natural systems, we need to find ways to transmit that knowledge.

For future generations to avoid some of the problems we are struggling to solve today, they must continue making better and more sustainable choices. This message is passed on from parents, through the media and increasingly at school: UNESCO declared 2005-2014 the "Decade of Education for Sustainable Development". Nations from Australia to France, from Chile to China, have included environmental concepts in their national curricula and are building eco-schools to ensure that younger generations meet the future with some of the tools they need to carry these ideas forward. But teaching children the complex concepts of sustainable development – interdependency, interdisciplinary thinking, intergenerational needs – is proving far more difficult. Curricula approaches for teaching sustainable development are still at an early stage.

We owe it to future generations to put mechanisms in place to achieve sustainability: nations can start enforcing emissions quotas and trading, to make each ton of CO_2 more expensive to emit, while at the same t ime developing and sharing alternatives for energy. It is important not to forget that we also owe it to ourselves: many of the consequences may seem like they are in the indefinite future, but that may be much sooner than we would like. Ageing populations, increasing poverty, stronger hurricanes, more frequent heat waves, increased flooding… the evidence suggests that these are no longer projections: the future is now.

Production and Consumption

In Samuel Beckett's novel *Malone Dies,* the main character decides to make a list. Taken literally, Malone's idea may seem like a typical example of the absurd. Yet so often with Beckett, an everyday occurrence can suddenly reveal unexpected depths, complexity and connections among what we are, what we do and, in this case, what we have. Imagine trying to write down everything you possess - every single thing. It might take a while, right? Yet go back a few generations and the problem would probably have been a lot easier for most members of your family - food, working clothes, maybe a set of formal clothes that lasted a lifetime, some household utensils and perhaps a few other goods. And that's all.

Even now, the 40% of the world's population living on less than two dollars a day wouldn't need much time to draw up their list of possessions. In OECD countries however, the economic expansion and social reforms of the past few decades have made the material conditions of most people's lives unimaginably superior to those at any other time in history. Rapidly developing economies such as China, India and Brazil are catching up and their consumption patterns are converging with those of OECD countries. Worldwide, more and more people possess more and more things. This has obvious implications for sustainable development. The billions of goods and components humans now own all have to be manufactured, transported and, sooner or later, disposed of. Consumption and production touch virtually every aspect of our lives: international trade, agriculture, energy, working conditions, social life and well-being. In fact, all the areas considered important to sustainable development actually have something to do with what producers bring to the market and what consumers - whether individuals, groups or governments - take from it.

The Material Society

We live in a "productivist" society, where growth and economic activity have long been the central focus of the activities we undertake as individuals and communities. World GDP has grown from around $16 trillion in the mid 1970s to over $40 trillion today. Companies are churning out more of everything and inventing new products all the time.

To take a simple example, let's go back to Malone for minute. He doesn't get very far with his list, overwhelmed as he is by a pencil and a notebook. Even such small and seemingly innocuous objects can give us pause for thought once we begin to add up the totality of their "weight" in the world. Every year, the Faber-Castell company alone produces 2 billion pencils, enough to reach from here to the Moon if laid end to end. An ordinary graphite pencil can write around 45 000 words, that's around 70 closely-written pages, or a line almost 60 km long. So Faber-Castell could probably meet the world's pencil needs for some time to come with a year's output. A quick look at any stationery store tells a different story. The modern marketplace offers an enormous range and quantity of even the simplest products. And manufacturers are continually trying to produce the next big thing, the next hot item that everyone will want. Workers, research, raw materials, machines, components, marketing and distribution, and numerous other services are mobilised to meet our demand as consumers for new and better products.

Although poverty and deprivation still exist, most people in OECD countries enjoy a standard of living that allows them to spend a significant share of their income on goods and services other than food, shelter, clothing or other basics. Even for the basics many of us can spend much more than is necessary for our physical well-being. Consuming is a pervasive fact of life and begins even before babies are born, when the parents' friends and relatives celebrate the big event with a gift. Babies themselves begin to consume, or to influence purchasing decisions, as soon as they can point at a toy or cereal box. In the US for example, discretionary spending by children aged 3 to 11 is expected to grow from $18 billion in 2005 to over $21 billion by 2010, while families will spend over $140 billion on consumer goods for their kids by 2010.

What Happened to the Paperless Office?

Indeed, consumption often seems to be the major criterion in defining a ctivities or social groups. A s the chairman of the IFPI, trade association for the recording industry, explained to a trade show in 2005: "A new generation has defined new ways of consuming music". Not "listening to" or "enjoying", but "consuming". In the past few decades, the technology for "consuming" music has undergone several major transformations: records, tapes, CDs and now the immaterial and intangible digital file.

Production has a far greater impact on sustainability than consumption, so taken in isolation, the fact that goods can now be obtained in a digital format is a good thing for sustainability. Selling a million copies of a song via Internet downloads saves tons of plastic, tons of packaging materials and tons of fuel to get the CDs to the shelves and the fans to the store. But, once again, we have to keep in mind that sustainability is not about taking things in isolation, but instead about examining the trends and interactions that make up the whole cycle of production and consumption. In this case, it means remembering that the virtual economy has physical foundations and that the digital product uses resources and creates waste. Over 7 million tons of phones, computers and TVs were sold in 2006, and this is expected to rise to almost 10 million tons a year by 2016. The servers that store all this information are using significant amounts of electricity – over 1% of the world's total.

The digital revolution has added hundreds of new objects to the market and often without the savings in r esources that i t was a ssumed the innovations would generate. People have been predicting for the past 30 years that the PC and other advances in electronic equipment would reduce the amount of paper used, leading to the "paperless office". In reality, consumption of paper products has almost tripled since the mid-1970s. Of course not all of this is due to office applications, but the introduction of e-mail into organisations for instance increased paper use by 40%. Other office technologies also have significant sustainability impacts, as in the 3.3 litres of oil it takes to produce a laser printer cartridge. And in spite of the possibilities for savings in travel through telecommuting, the vast majority of people still work in an office, with fewer than 2% of workers working more than eight hours a week at home.

Why has technological progress and the so-called information society not produced the resource (and time) savings that should be possible? Well, for a start, goods have become cheaper – you can now buy a laser printer for the price you would have paid for the cheapest inkjet printers five years ago – and world living standards are rising, increasing the number of buyers of every kind of object. The answer to this question also has to account for how people use the technologies, favouring throw-away objects rather than reusable ones for example. Making production and consumption sustainable means considering the whole life cycle of a product, from the raw materials needed for production to labour costs and conditions, to the costs of transport, retail distribution, use and waste disposal.

Production and Consumption

Production and consumption together form the backbone of the economy. They also help to determine social status and shape the natural environment. We can better understand some of these issues by looking at an everyday object, the mobile phone.

Thirty years ago, the idea of a tiny radiotelephone capable of calling practically anywhere in the world was the stuff of futurist fancy. To day, not to have a mobile is to pass for an oddball, or a technophobe. Even in countries where income is very low and poverty a major concern, mobile technology is relatively common, having leapfrogged traditional telecoms in many cases. There are only about 14 fixed-line telephone subscriptions for every 100 people in developing countries, but over 33 mobile subscribers. And the trend for mobiles is moving sharply upward, while that for fixed is actually declining in the developed countries according to the International Telecommunications Union. What does this mean for sustainable development? It means that more people than ever have access to modern communications networks and the benefits they bring. As Internet via mobile phone expands, it will mean that people who can't afford a computer can access the Web. It means that banking services can be made available without having to build banks. But since we are looking at all that goes into (and comes out of) a product, we have to examine the physical impact of all these phones, too. Worldwide mobile subscriptions had reached 3.3 billion by the end of 2007, and a billion mobile phones are sold each year. An average user changes phones every 18 months to two years and very few of the old ones are recycled. Although one phone may not make much difference, the life cycle of billions of phones is a major issue.

Out of Sight, Out of Mind

What really happens to all that waste at the end of a product life? What does it m ean exactly to "store or reuse" it? Where does it really go when "exported"? In 2006, the tanker *Probo Koala* offloaded a cargo of toxic wastes onto trucks in Abidjan, the Ivory Coast capital. The trucks then dumped the waste at 14 municipal dumps around the city. The resulting pollution killed at least 7 people and poisoned 9 000 others, provoking vomiting, nosebleeds, headache and rashes. The story starts in Amsterdam, where the cost for treatment would have been •500 000. The ship sailed on to Estonia, which refused to let the waste enter its territory. It was then sent it to Africa, and a newly registered company was paid $18 500 to dispose of the waste. The *Probo Koala* case is only one example of the "grey areas" involved in disposing of material waste. Like similar cases, it reveals some of the many governance, regulatory and even geopolitical factors that can impede or hinder sustainability.

The raw materials side of a product's life cycle can also have a significant influence on people's quality of life, their health and safety. This can even inadvertently contribute to conflict, as in the case of the capacitors found in phones, laptops and other electronic devices, which use a substance called tantalum, valued for its good thermal conductivity and energy efficiency. Although Australia is the world's biggest producer, the increased demand has made other sources attractive, too. A UN report revealed that the civil war in the Democratic Republic of Congo was being partly financed by illegal mining and trading of coltan, the African abbreviation for columbo-tantalite, a source of tantalum. Phone manufacturers do not buy coltan directly, and knowing which components suppliers are using illegal materials is extremely difficult, the more so given the sheer number of components that go into a phone – 500 to 1 000 depending on the model.

Getting the Price Right

If asked to describe how the price of goods is fixed, most of us could describe the various factors taken into account, such as raw materials, labour, profit margins and so on. But these economic parameters do not tell the whole story. The *Economist* magazine has devised an amusing and instructive index it calls the Big Mac index to compare prices around the world. For example, an average American would have to work for around ten minutes to buy a hamburger, while a Kenyan would have to work for three hours. We can use this novel way of looking at things to think about production and consumption.

Most of us wouldn't mind working 10 or 15 minutes to buy a hamburger. But what if you had to find the 2 4 00 litres of water needed to make it? You are probably familiar with the idea of carbon footprints – the amount of CO_2 generated by various activities such as travel. The water footprint is a similar figure calculated for the use of fresh water. It is based on the idea of "virtual water". A tee-shirt for example contains no water, but it takes 11 000 litres on average to produce a kilogram of the cotton it's made of, once you include irrigation, bleaching, dyeing and all the other steps in the production chain. Virtual water is therefore the total amount used to make a product, adding up to 2 7 00 litres for a shirt. Unlike the fuel that produces CO_2 though, water is rarely sold at a price to the main users that takes into account all the costs incurred. Often the costs of providing infrastructure, purification, waste treatment and distribution are subsidised, so there is less incentive to use water sustainably.

Positive and Negative Aspects

Economists use the term externalities to describe the positive and negative aspects not accounted for directly in prices. Bees are often cited as a typical positive externality. The beekeeper raises them to be able to sell honey, but they pollinate all the plants in the area, providing a benefit to farmers and gardeners. Pollution from factories is a typical negative externality where the cost to public health is not included in the productions costs the polluters pay. Getting the price right means getting closer to the "real" price – and this requires factoring in what the production and the consumption of something will generate in terms of externalities.

Assigning a value to things that previously fell outside systems of accounting, budgeting and measurement is a major challenge. It is not easy, for example, to assign a value to natural resources. For some, such as forests, we can calculate the value of what is produced because i t i s bought and sold and therefore has a monetary value. Still, knowing the price of wood – and therefore knowing something about what a forest is worth monetarily – doesn't tell us anything about its value in offsetting CO_2 emissions, its role in preserving biodiversity or its spiritual and cultural value to people whose way of life depends on it. Assigning a value to clean air is harder still. Air pollution generates costs in the form of increased rates of disease, lower real estate values in "dirty" areas, and current and future damages related to climate change. Clean air is worth something to us, but can we say how much exactly? The "ecological services" concept sees that putting a price on these services is a useful way for them to be valued – getting the "real cost" of resource consumption into the equation.

In the meantime, nearly two-thirds of the services provided by nature to humankind were found to be in decline worldwide by the UN's Millennium Ecosystem Assessment. The UN points out that the costs may be borne by people far away from those enjoying the benefits, as when shrimp eaten in Europe is farmed in a South Asian pond built in place of mangrove swamps – weakening a natural barrier to the sea and making coastal communities more vulnerable.

Current trends in global production and consumption patterns are unlikely to change significantly, meaning that externalities will increase. Goods are becoming cheaper and are being transported in ever-bigger amounts from one side of the world to the other. Even a simple pot of yoghurt may have travelled over 3 000 km by the time it reaches the table and require inputs from several countries for its ingredients, production and packaging. W e r e place goods much more quickly than in the past. Few people still use an MP3 player they bought five years ago, for instance, but an old gramophone would have lasted for decades. And unlike our grandparents' generation, we throw out rather than repair everything from vacuum cleaners to socks.

Technology might reduce some of the negative impacts on sustainability of production and consumption but it will create others, and technological improvements are often outpaced by growth in consumption. Cars are now much more fuel-efficient than before, for instance, but air pollution is getting worse because so many more people have cars. While market-related transactions have grown ever more efficient and many private goods such as food, cars, air-conditioning and designer clothes can, in principle, be afforded by anyone who would like them, the growing externalities of these transactions has made many "public" goods increasingly scarce: clean air, silence, clear space, clean water, splendid views, and wildlife diversity are highly valued and sought after. Nearly every transaction of private goods carries an invisible cost, paid by everyone through degraded public goods. Achieving "decoupling" between continued economic growth and prosperity and the negative externalities created by such development is therefore a major challenge for achieving "development that lasts".

The way products are produced and consumed affects not just the environment, but also living and working conditions. Once again, looking at an everyday object can help us understand the issues. Fred Pearce, *New Scientist's* senior environment correspondent, set out to discover where the cheap pair of jeans he bought in London came from. H is investigation took him to Dhaka, the capital of Bangladesh, where hundreds of thousands of women make clothes for the big brands of Europe and North America – for just under one dollar for a ten-hour day. But the women point out that the sociologists who conduct the regular "social audits" of factory conditions do not learn the truth: before they come, "the managers instruct us what to say about working hours and holidays and conditions".

Does this sound unfair? It might come as a surprise to learn that none of the women Pearce spoke to supported boycotts of the goods they make. As he points out, they are the first w omen in conservative, rural Bangladesh to have any sort of freedoms beyond those allowed by their husbands i n the villages. "The garments industry has created a revolution in women's economic empowerment," says Mashuda Khatun Shefali, who runs an NGO that supports women garment workers and tries to improve their conditions. Another campaigner, Nazma Akter, notes that, poor as most were, "women are becoming an economic force here. This is the first time they have had jobs. They are independent now. They can come and go; nobody stops them. Don't take that away from them."

What these women ask from consumers in the rich countries is to pay a fair price for the goods they make and not to ask them to sacrifice their health and well-being for a marginal improvement in our standard of living. Women's status, workers' rights and fairer trade are as much a part of sustainable development as protecting the environment. Making production and consumption sustainable implies recognising the true costs of what w e make and w hat w e buy, across the entire supply chain, from sourcing to retail distribution to waste disposal.

Connecting the Dots

What can be done concretely to promote sustainable production and consumption? Even quite small actions can have a large cumulative impact. According to the US Environmental Protection Agency, if every home in America replaced just one incandescent light bulb with a compact fluorescent one, in one year enough energy would be saved to light more than 3 million homes and prevent greenhouse gas emissions equivalent to those of more than 800 000 cars.

Or, as the European Lamp Companies Federation calculated, if Europeans switched t o the more efficient bulbs, the continent would need 27 fewer power plants. This trade association has launched an effort to encourage European consumers to switch to the energy-saving bulb. Australia, Japan and the state of California plan to implement bans on incandescents, seeing the need for government to step in and make the switch more rapid and definitive. The European Commission is talking about a phase-out, too.

Most successful efforts at influencing markets involve co-ordinated efforts – where producers, consumers and governments each have a role to play. Only through the combination of these efforts can old behaviours and processes be transformed on a large enough scale to make sustainable development the rule rather than the exception.

Co-ordinating our Efforts

How can producers integrate sustainability into product design, manufacture and distribution without sacrificing traditional factors such as profit or brand image? We can't really be sustainable consumers without sustainable product choices.

Design is arguably the most influential single stage in the process, since it determines the rest. First of all, the design has to consider the product as part of a product system and consumer lifestyle. If a phone integrates a camera and an MP3 player, will users then buy only the phone, saving the environmental and other costs of manufacturing three separate products? Or will they also buy the others, adding to the burden? Design also means choosing the materials from which the product will be made. Can the product be designed in such a way as to maintain the required physical properties while using fewer materials? Could it be made of renewable, recyclable materials? Design also influences manufacturing by determining the number of steps in the production process.

Questions specific to manufacturing can involve how to reduce energy consumption or pollution, or how to improve worker safety. Once the product is made, what is the most sustainable way to package it for shipping and sale? Should it be transported by air, sea or land, by rail or truck?

Design and manufacturing play a big part in determining how long something will last, how many other products it will need to work (*e.g.* batteries) and if it can be repaired or maintained. And finally, producers also have to think about what happens to the product at the end of its cycle. Once again, good design can make a difference. Are the materials used easy and safe to recycle for example? Can parts of the product be refurbished and reused?

Think of your first major purchase – maybe your first bike or a car. Think of all the factors that went into that decision. Cost was probably one of them but so were colour, texture, brand, and a whole host of aesthetic and emotional associations that attract us to the objects we buy. Businesses spend a lot of money on market research and advertising (over $650 billion a year worldwide on marketing alone), trying to understand those associations, to be able to predict and influence people's tastes and preferences. At first glance, the pervasive influence of marketing and the push to consume seems to be at odds with sustainable consumption after all, unchecked consumption has played a big part in creating many of the problems we currently face. And yet, if we don't know of a less-polluting product for cleaning the kitchen sink or the difference in energy use for an appliance, then we can't make a better decision. Including a sustainability dimension in marketing and distribution allows us to add another, very important criterion to purchasing decisions. It provides information to consumers and is of course a means to influence them in a "sustainable" direction.

Experience of the past decades has shown that providing sustainable products to the niche market of "green" or "fairtrade" consumers is not enough to change patterns on a larger scale, although it has been a significant factor in pushing both producers and consumers in a new direction. The proliferation of eco and fairtrade labels in the last ten years is evidence of that evolution. For a long time, products such as organics and fairtrade suffered from their reputation as "specialty" items, attractive only to a small category of consumers willing to pay extra to consume according to their beliefs. But this trend has started to change, for several reasons.

For one, more people are aware of the impact that their choices have on the world around them. Problems related to unsustainable consumption – the cost of petrol, for example – have become more "real" and have begun to touch the average consumer in more concrete ways. As a result, a growing number of consumers are beginning to ask important questions about what they buy; how much waste is created by the product and its packaging; how much water, energy and other resources go into its production (and into its disposal); and what are the living and working conditions of the people who produce the goods.

Second, the products themselves have become more mainstream. A wide range of companies are using the tools of marketing to give products and services considered sustainable a cool or sexy identity. Manufacturers are designing more products that are appealing for their aesthetic qualities or their ease of use *as well as* for their environmental and social sustainability – a look at the latest fuel-efficient car designs provides a good example. Most major retail grocery chains now offer a selection of certified "environmentally friendly" products whose market shares are increasing: for instance organic coffee imports to North America grew by 29% over 2006-2007 compared with 2% for conventional coffee.

The changes in consumer awareness and the proliferation of more sustainable products and services seen over the past few years are encouraging. Some critics and consumer advocates rightly point out that some of this is "fluff" or "greenwashing". Products that claim to be environmentally friendly can look much less so once you take a hard look at the list of ingredients or analyse the entire product life cycle. Buying fairtrade coffee cannot alone solve the problem of poverty. While this may be true, it doesn't detract from the fact that sustainable consumption and production are critical to the success of any kind of sustainable development. The fact that more people and businesses recognise and even wish to capitalise on this realisation can be seen as testament to the growing mass appeal of sustainability. Efforts at making the consumer society sustainable are gaining momentum.

Good Production and Consumption

Finding the right policy tools to encourage good production and consumption practices and avoid overlap and inconsistency is one of the biggest challenges that governments face. Elected officials are under pressure from constituents and special interest groups to respond to issues perceived as important and follow a given line of action. For one, decisions must be based on thorough research and sound evidence, or they risk not solving the problems at all and often create additional ones. Governments have the enormous advantage however of being able to make laws and impose regulations. One solution at their disposal is simply to outlaw products and behaviours that are seen to be doing more harm than good. This is what happened to CFCs (gases used in refrigerators and aerosol sprays) that were damaging the ozone layer. The *Montreal Protocol on Substances that Deplete the Ozone Layer* came into f orce in 1989, and 191 countries have now signed it. Since then, atmospheric concentrations of the most important CFCs and other related gases have either levelled off or decreased. The plastic shopping bag is another example. Bangladesh banned them in 2002 following a movement that began in the 1980s in Dhaka. Discarded bags were blocking drains during the monsoon, causing flooding.

Critics of such plans point out that packaging is a much bigger source of waste than plastic bags. Retailers should be using their buying power to influence packaging choices from their suppliers and governments should be using their regulatory powers, rather than passing off the responsibility for reducing waste to consumers alone. This kind of argument is heard often in discussions around sustainable development: who is primarily responsible for changing bad practices? Where is the best place to focus our efforts? One might counter that, in the above case, why not do it all? Encourage customers to reduce needless use of plastic bags by charging for them, a m easure that has shown consistent results. A t the same time, encourage retailers to use their influence to reduce packaging and to implement any other measures to curb resource use and waste. And develop government standards for sustainable and recyclable packaging. The town of Modbury in England shows what could be done. The town became the first in Europe to ban plastic bags after shopkeepers agreed to the move following a few weeks campaigning initiated by Rebecca Hosking in the pub one evening. And as she explained to the *Guardian* newspaper, Modbury is a town "... that's always been very conservative. If we've done it, it proves you don't have to be one of those 'green' towns to change over."

The fact is that often, although we might see what would help, changing things is a different story. As former UK Environment minister Margaret Beckett points out, "*Many unsustainable behaviours are locked-in and made 'normal', n ot just by the way that we produce and consume, but by the absence of easy alternatives*". Creating those alternatives is thus a priority, one that can benefit from the force of government intervention. The European Union launched an Integrated Product Policy pilot project involving m obile phone producers, component manufacturers, telecoms operators, consumer groups, recyclers, NGOs, government representatives and researchers. A number of governments are making efforts to promote recycling through initiatives such as France's "éco-participation", a surcharge on electronics goods of •0.52 per kilo that goes towards paying recycling costs.

As the CFC example suggests, persuading producers and consumers to change is not always the most efficient way to tackle the issues, nor is it enough to produce a big enough change on a large enough scale. The individual producer or consumer generally has little power to change things or interest in doing so. A manufacturer who decided to go it alone and implement stricter environmental standards or m uch better pay and working conditions than competitors would be at a considerable disadvantage. People have a hard time reducing their consumption even of products they know are damaging their own health or adopting behaviours that would improve it. Although consciousness about sustainability is improving, waiting for a change of attitude is not the best policy.

Powerful Persuaders

Governments have powerful persuaders at their disposal: regulations and taxes. These have long been applied to economic and social policy. Can they work for sustainable development? The answer is yes. Governments originally tried persuasion and communications campaigns to get consumers to use less energy in their homes. They worked with producers to put labels on appliances indicating their relative energy consumption. In the end, they enacted minimum energy efficiency standards, which forced companies to change the design of their products. Although consumers can have some influence, regulating processes and products is the fastest way to sustainable production.

Experience also shows that environmentally related taxes ("green" or "ecotaxes") and emissions trading can be efficient instruments. They can force polluters (whether producers or consumers) to take into account the costs of pollution and can help to reduce the demand for harmful products. Ireland's 2002 "plastax" led to a 90% reduction in the use of plastic bags. The 1990 the United States Clean Air Act pioneered emissions trading. It included a requirement for a major reduction in sulphur dioxide (SO_2) and nitrogen oxides by 2010. Each polluter had the "right" to emit a certain quantity of SO_2. If they managed to emit less they could keep their permit or sell it; if they emitted more they had to buy quotas from another polluter (or pay a fine greater than the cost of the quota). By controlling the number of quotas allotted, the Environmental Protection Agency has already brought emissions down 50%. Other countries have since started emissions trading schemes.

All OECD member countries now apply several environmentally related taxes (375 in all plus around 250 environmentally related fees and charges). The taxes raise revenues of about 2% to 2.5% of GDP, with 90% of this revenue from taxes on motor vehicle fuels and motor vehicles.

The environmental effectiveness and economic efficiency of green taxes could be improved if existing exemptions and other special provisions were scaled back or if the rates were made highly dissuasive. The rise in fuel prices shows that cost can be a big factor in changing behaviour, but the level of carbon taxes is generally too low to make it worthwhile for manufacturers to change production methods. Moreover, higher tax rates can face political opposition for two reasons. First is the fear of reduced international competitiveness in the most polluting sectors of the economy. This is why the taxes are levied almost exclusively on households and the transport sector, leaving energy-intensive industries totally or partially exempt. Second, exemptions create inefficiencies in pollution abatement and are contrary to the OECD's "polluter pays" principle.

Another point to bear in mind i s that, in practice, environmentally related taxes are seldom used in complete isolation. A labelling system for instance can help increase the effectiveness of a tax by providing better information to the users. Combining a tax on energy use with subsidies or government standards for better isolation of buildings can be a good way to encourage energy savings. The combination of a tax and a voluntary approach can increase the political acceptability of the tax, although this may reduce environmental effectiveness or increase the economic burdens placed on other groups.

What's Next?

Seeing, calculating and understanding exactly what we buy, use and waste is just a beginning. It's an important first step in gaining awareness, but it would be unfortunate and ineffective if tools like the footprint are used only as gadgets. Realising that it takes x amount of water or oil to make a given thing has to lead us to at least two questions:

- How could we develop more efficient processes that are implemented on a large scale - in order to significantly reduce resource use and the negative impacts of production?
- % How should we use the role of "informed consumers" to make sustainable consumption decisions on a scale that produces real results?

Production and consumption are at the heart of sustainability. From a material point of view, life today is far better than it was a century ago for most people. If the improvement is to continue and benefit the world's population as a whole, the economic, social and environmental pillars of sustainability will have to be reconciled. This will not be easy. Sometimes what might favour one pillar will damage another. Different social groups will have different priorities and projects. But debating these differences and finding solutions is not beyond us, in fact it's what democracy does.

Measuring Sustainability

The previous pages explored what sustainable development means and how contemporary societies are trying to implement it. But how do we know if what we're doing is helping or hurting, or having no effect at all? How do we know that one way of doing or making things is more sustainable than another? That a city, region or country is doing well in terms of sustainable development? How do we calculate today's needs and measure our progress in meeting them? And how can we get some idea of how our decisions will affect the future our own and our children's? To be able to answer these questions, we have to decide first on the basics. What is important to us? What resources do we need to keep track of? What are the different factors that contribute to our quality of life and well-being?

We often have to do this kind of calculation in our daily lives. We know how much money we have and how much we need to buy food and pay the bills. We know, albeit imperfectly, what expenses we'll have to meet later, and we know that there will probably be some unexpected ones, too. We know what we'd like to do and if there is something left over we can spend it on dinner and a movie, clothes or maybe even a holiday. All this depends on counting, planning and reasonable guesses. It is based on measuring our "resources" and following what happens to them, establishing priorities among all the things we have to do and those we'd like to do, sometimes leading to difficult choices. To put it another way, we all have an information system (however informal) to let us see how we're doing just now, make predictions about what we will be able to do in the future, and monitor whether or not we are living within our means.

What makes a good and sustainable society and how can we judge our progress towards creating one? It's more than money, to be sure. The answer lies in a range of factors that make significant, often essential contributions to our "success" from access to education, health care and functioning ecosystems to freedom, justice and cultural expression. Developing and refining accurate measures of these things w ill a llow us to build a more sophisticated and stronger knowledge base, and potentially to speed up progress towards achieving them.

What should we Count and When?

Agreeing on the best indicators to measure sustainability or progress towards sustainable development is a challenge. An indicator is a summary measure that provides information on the state of, or change in, a system. Indicators give us a snapshot of how we are doing at a given point in time relative to what we've decided is important. Indicators also provide feedback on the effects of our actions and government policies. And indicators have to be able to adapt to the changing conditions and the content of policy.

At first sight, measuring sustainable development seems impossible, the subject is so vast and the influences so many -climate change and child care, business ethics, government policy and consumer trends to name but a few. We know that sustainable development involves economic, social and environmental variables - all of which must be measured to some extent. As shown in the annual *OECD Factbook,* there exist a wealth of indicators from traditional macroeconomic measures, such as gross national product (GNP) and productivity, to environmental indicators, such as water consumption and polluting emissions, and social statistics, such as life expectancy and educational attainment. But which indicators are the most important to sustainable development?

The issue is made even more difficult by the fact that as well as being multidimensional, sustainable development is a dynamic concept. Quantifying it requires juggling a number of parameters including time horizons. Economic, social and environmental phenomena operate at different rhythms to each other (and even within any one of these, several time scales may be operating simultaneously). For instance, the legal systems of most countries are still strongly marked by codes dating from the time of Augustus and the Roman Empire, and their basic principles change slowly. Some technologies, on the other hand, change at a rapid pace your new computer is probably out of date before it's out of the box.

Consider the economy: if you're planning a major energy project, you have to think at least 50 years ahead, but if you're trading on financial markets, the nanoseconds it takes price data to go from one exchange to another can mean substantial gains or losses. For its part, the environment shows how the pace of change can suddenly accelerate, as when fish stocks rapidly disappear after declining slowly for years.

Moreover, we have to bear in mind that sustainable development is a process linking what happened in the past to what we're doing now, which in turn influences the options and outcomes of the future. In a sense, it's like walking along a path backwards – you can see where you've come from, you can see more or less where you are, and you can get a rough idea of where you're heading. But you can't tell if the other paths branching off from your route are dead ends, short cuts or ultimately heading in a completely new direction. Likewise, it's difficult, if not impossible, to say if any given point along the development path is sustainable, it may be more or less so depending on how far we've come, what happens next, which new perspectives open up, and how attitudes and other influences change.

These uncertainties complicate sustainable development measurement. And developing measures is not a purely statistical or technical exercise. It touches on two very sensitive areas for all societies: government accountability and social participation. Measuring progress on sustainable development (or any other important area of policy) with reliable information is a key ingredient of the democratic process. It makes governments more accountable and gives people a tool to participate more actively in defining and assessing policy goals.

Measuring Progress

Progress has long been defined and measured in purely economic terms. A country's overall performance or well-being is often expressed in a kind of shorthand fashion through one "superstar" measure: gross domestic product. GDP calculates in monetary terms the value of what counts as production. You may also be familiar with GDP per capita, how much each person in the country would have if everybody got an equal share of GDP. While these measurements may sound fairly straightforward, they omit important factors and include others that we would probably rather live without.

For instance, software that is sold goes into the calculation, but freeware does not. On the other hand, spending on cleaning up an oil spill is included in GDP. In this case, not only is something negative being counted as contribution to "production", but substantial costs to well-being are left out of the equation and remain invisible. Housework, caring for one's children and volunteer work are not counted, though they contribute value to the economy and our daily lives. Per capita GDP is a fairly crude measure, too. As an average, it does not address issues of distribution benefits of economic productivity may go disproportionately to only a small percentage of a population, even though the average looks good.

Such indicators are useful for giving a rough idea of what's happening in the economy and comparing national performance. But there is a large and growing gap between what official statistics, like GDP, tell us about "progress" and what people experience and care about in their daily lives purchasing power, public services, quality of life and so on. The notion that the three pillars of sustainable development should be considered equally important, interconnected and interdependent reflects the idea that economic progress alone is not enough to guarantee that a society is "headed in the right direction". Other factors, such as access to good health care and education, can be equally or more important to creating well-being, life satisfaction and health over the long term, both for the current and the future generations.

Individual calculations can help us gain awareness of the progress we are making in terms of income or health for example. But if these calculations don't fit within some kind of framework that structures the analysis, we can't get any sense of how we are doing or where we are headed. We need something that allows us to understand the big picture emerging from the data and use it as the basis for policy and action. A conceptual framework can help us select those indicators which best measure what we want to assess in a coherent and consistent manner. It will also help compare sustainability across different societal levels.

Finding ways to make comparisons between countries or regions who may not share the same history, culture, level of economic and social development or physical conditions is a challenge. Reaching this goal requires an ongoing dialogue on needs, resources and how they are evolving, as well as a flexible approach to building sets of indicators that can supply the most useful evidence and information. Individual indicators are the basic building blocks of this process.

The task is complicated further by the fact that w hat is considered important to sustainability varies to some degree from place to place the quality of water, land or air; people's income or access to energy; life expectancy or any number of other indicators. How then do we develop methods for measurement that reflect a particular context or geography and also allow us to work across institutions and geographical boundaries to advance sustainability worldwide?

A range of indicators can be used to compare the relative situations of countries, assess their strengths and weaknesses and identify domains where policy intervention is required. It would be much easier to have one list of indicators for everyone, allowing quick comparisons among different places and across time, but this is not so simple: what matters in California is not an exact match to what matters in Helsinki or Bangalore.

And yet, common indicators are needed if countries or localities want to compare their progress on sustainable development with others. From this, they can learn what works and what doesn't work. This is why, as measurement of sustainable development has evolved, many local, national and supra-national bodies such as the United Nations and the European Union have developed and refined sets of indicators. Along with international organisations and NGOs, they have put considerable effort into discussing and tuning their indicator sets to improve measurement of sustainable development and allow comparisons among countries or other levels of administration.

The Capital Approach

The key idea of sustainable development is the linkage between the well-being of the current generation and the well-being of future generations. To make this connection we can use the concept of capital. In economic terms, capital is a stock that is used in production over several years: think of a machine or a factory. Capital can be created by investment, and it is consumed over years and eventually wears out. The concept of capital can also be applied to sustainability, allowing us to measure all types of wealth that contribute to well-being more comprehensively. Economists use the concept of *national wealth* to indicate this broader measure.

The "Capital Approach" is a framework for measuring sustainable development which operates on the principle that sustaining well-being over time requires ensuring that we replace or conserve wealth in its different components. It emphasises the need to focus on the long-term determinants of development not to the exclusion of current needs, but rather according to a principle of sustainability: development that can be continued into the future. This approach allows us to discuss and evaluate how what we do now will work in the very near, medium and long-term, and how to talk about whether or not there is "progress", "regression" or "stagnation".

With this model, a society's total capital base encompasses five individual types:

- *financial capital* like stocks, bonds and currency deposits;
- *produced capital* like machinery, buildings, telecommunications and other types of infrastructure;
- *natural capital* in the form of natural resources, land and ecosystems providing services like waste absorption;
- *human capital* in the form of an educated and healthy workforce;
- *social capital* in the form of social networks and institutions.

Conceiving these different forms of capital as inputs into the production of well-being allows us to calculate national wealth as the sum of the different kinds of capital.

Sustainable development requires making sure that national wealth per capita does not decline over time and, when possible, that it increases. For example, if we consume all our natural capital and do nothing to preserve or increase it, this source of well-being will dry up, leading to unsustainable outcomes. The capital approach allows monitoring that capital stocks are not "drawn down" too low. Norway's management of its petroleum stocks provides a good example. With some of the world's largest petroleum reserves, Norway could spend the profits from the sale of its petroleum on any number of programmes. Instead, Norway invests these profits to be sure that when the oil reserves are depleted, other sources of income will be in place. To put it another way, just as financiers seek to maximise their capital base and the dividends it produces, we should maximise the financial, produced, human, social and natural capital base of our welfare and make sure it continues to pay dividends in terms of well-being over time.

This sounds very straightforward, but "maximising" capital involves making important decisions about what can be used up and what must be preserved. One important question: can the different types of capital be "substituted" for each other, as long as the total sum is maintained, or does each type have to be maintained at a certain minimum level? The practical answer to this question is that it depends on circumstances. In most circumstances, some specific categories of "critical capital" will be essential to the proper functioning of the world and our societies, things which perform essential functions and can be replaced at the margin only at huge costs.

A liveable climate is perhaps the most striking example it doesn't much matter what our national wealth adds up to, if climate change makes life on Earth or in certain parts of it impossible. Although environmental types of essential natural capital are the first to come to mind, aspects of social and human capital can also be critical. When the social networks and norms that are a basis for communities are depleted, societies break down, as in the case of conflict and war. Similarly, without education, human capital cannot be sustained, making overall sustainability impossible.

Global Dimension

That said, many of the key issues for sustainable development are transboundary and even global, meaning they have impacts beyond political or geographical frontiers. Environmental issues such as air pollution or biodiversity loss are obvious examples, but economic and social questions are increasingly globalised too, trade or migration being the most obvious examples. Whatever measurement framework is used, it will need indicators to reflect sustainability for a variety of specific contexts and others that capture issues of global scale, such as climate change.

The WWF uses an analogy that is worth bearing in mind when trying to understand what indicators are and how they can be used. Think of a car. Dials and other displays give drivers a range of indicators as to how the car is performing, but not all of this information is relevant at any given time or for a given purpose. The oil temperature might be perfect, but if you run out of fuel the car is going to stop anyway. And bad drivers will still be a danger, no matter how many fancy gadgets are on the dashboard. Sustainability indicators are like the car's instruments, addressing individual items (energy reserves would be a direct analogy) or combining indicators across a number of domains to give a fuller picture (just as how "good" a car is depends on fuel consumption, safety, comfort, etc.).

Many companies have developed their own metrics for assessing the economic, environmental and social impacts of their facilities and products. Some are combining these into composites or simple indices, which are more likely to get the attention of CEOs. Larger corporations are also formulating ways to assess the sustainability of their supply chains of smaller companies. Ford of Europe, for instance, uses a Product Sustainability Index as a management tool to assess the potential impacts of motor vehicles on a range of factors. This is an engineering approach which combines eight indicators reflecting environmental (global warming potential, materials use), social (mobility, capability, safety) and economic (lifecycle costs) vehicle attributes. Indicators are not aggregated into one ranking but rather tailored to the needs of various departments of the company.

Assessing Sustainability

Indicators and sets of indicators are the basis for assessing progress on sustainability. Many different assessment methodologies exist, for example: regulatory impact assessments, poverty impact assessments, environmental impact assessments and strategic environmental assessments. But in these approaches, the exercise tends to focus on a particular pillar of sustainability and economic aspects tend to dominate. W hat w e need are a ssessments that examine economic, environmental and social impacts and also the longer-term. In other words, we need sustainability impact assessments that can be applied to policies, programmes or agreements; to the national, regional or international levels; and to particular sectors of the economy.

Indicators and assessment tools already exist. The EU *Sustainability A-Test* site gives a good idea of the number of tools available. It presents 44 different types of tools for assessing sustainability classified into participatory processes, scenarios, multi-criteria analysis, cost-benefit analysis, accounting tools and models.

Whatever the methodology (indicators, models, surveys, cost-benefit analyses, cost-effectiveness studies), the procedures for conducting sustainability assessments have to be transparent and encourage the involvement of all concerned. The assessment has to be able to identify economic, environmental and social impacts but also the synergies and trade-offs across these dimensions. Different stages must be specified including a relevance test whether a sustainability a ssessment is even needed for the problem at hand.

Assessment results have to be presented to policy makers and others in clear and understandable terms. Even a well-designed assessment, carried out in a thorough manner, will have no influence if it neglects the political factors that impede its use. Most approaches may be too complex and too long for policy makers, while the existing bureaucracy may prefer traditional approaches rather than new assessment techniques. Moreover, sustainability assessments are often seen as an add-on rather than as an integral part of policy making. As a result, assessments may come too late with limited consideration of alternative policy options. Approaches for making better use of indicators and assessment tools are needed if we are to operationalise the concepts of sustainable development.

Good Life

In essence, sustainable development is a means for improving our quality of life today in ways that can be maintained over time. It teaches us to value all that contributes to our well-being, even if like ecosystems their "worth" cannot be easily calculated. Our job as citizens, scientists or policy makers is to think through the best ways of including what is crucial to our existence in the balance sheet. And to make decisions that keep us out of the red.

Sustainable development has heavily influenced the debate on how we – societies and governments conceive of our role in the search for better, more balanced ways of living. In doing so, it has given new life to a conversation dating at least as far back as Plato: What constitutes the good life? And how does one go about creating it? These seemingly straight forward questions are not simple to answer. Happiness, satisfaction, well-being, welfare and wellness we use all of these concepts to express the idea of what gives life quality, of what makes it good. The questions that we pose as individuals are, in large part,those that drive the debate at the group level. Thoroughly exploring what constitutes progress in these areas what the goals are, how far we are from reaching them, what kinds of trade-offs it will require to get there is the central task faced by citizens and governments. The tools and measures elaborated through sustainable development will continue to inform this exploration, providing a basis for the on-going work of improving the ways we govern and the ways we live.

Measurement of sustainable development helps us in two important tasks: evaluating where we are going and assessing the effects of specific policies, not only on the current generation, but also on future generations. One essential principle underlying any attempt at measurement is to understand what goes into the measurement process – what data are the most important, how they are collected, how they are compiled to provide evidence and how they can be expressed in different ways. For if we make ourselves smarter as an audience, then we can more easily select and understand the measurements w e need to make good decisions, for us and for future generations.

Government and Civil Society

In February 2008, rioters in Burkina-Faso took to the streets, angry at the jump in food and fuel prices over the past year. They burned petrol stations, trashed government buildings and stoned a government delegation that had come to discuss the problem. Within a few weeks, similar scenes were repeated in over 30 countries around the world, from Haiti to Somalia, Yemen to Indonesia. The world's poorest were not the only ones feeling the pinch. Italians and Mexicans were also up in arms over the cost of pasta and tortillas, whose price has considerable symbolic value. In the year leading up to the crisis, prices for many staple foods, including wheat and rice, doubled or even quadrupled. The consequences were visible to consumers on shop shelves worldwide, and the effects ranged from the plummeting popularity of governments to the riots described above.

The interaction of economic, social and environmental factors produced the crisis. What are these factors? As the world economy has expanded, prices of all commodities have increased. Higher standards of living have driven up demand for beef and dairy products, and added to the energy needs of modern agriculture, already a big consumer of oil and other petroleum products for pesticides, fertilisers and transport. Planting crops for biofuel intended to reduce dependency on oil has taken land away from food production, tightening supply and driving prices higher. Major food producers, including Australia and Myanmar, have been hit by droughts and cyclones respectively, further limiting supply. Changes in international trade have led some countries to rely on imports, whose prices they can no longer afford.

Making Changes

At its most basic, politics is about making decisions on what is important to a society and how these important issues should be handled. It is a process by which people and groups who may not agree attempt to translate their beliefs into workable rules, or laws, to regulate life within a community. The structures of government that manage these processes are often conservative and the impetus for new thinking has often come from outside. In many cases of major societal change, the pressure to transform laws and attitudes has come from visionary individuals and groups, or "civil society organisations" arguing their case until a critical mass of public opinion and political backing has been reached. Then, what was new and at times shocking, irritating or seemingly impossible, became the norm a part of our political and social fabric.

Think of the changes developed countries have seen over the past 100 years. At the start of the 20th century, horses and walking were the main means of transport, even in the rich metropolises like Berlin, London or New York. If aa street was lit at all, it was probably by gaslight. Before penicillin, infectious diseases were often fatal. Women were killed in the fight for the right to vote. Go back a few decades more and slavery was considered normal. Children under 10 years old worked 12-hour shifts in factories, as they still do in some countries today.

How did conditions and attitudes change? How was what seemed natural and unchangeable swept away? There's no single cause for the major changes in human history. Visionary individuals argued and organised for change. Sometimes a book or other cultural event provoked a shift in conventional thinking – Dickens' *Oliver Twist* cast a harsh light on England's 1834 Poor Law Amendment, while Upton Sinclair's novel *The Jungle*, published in 1906, showed the appalling working and sanitary conditions in the meat industry contributing directly to the creation of the US Food and Drug Administration.

What can this teach us about improving the world? About increasing well-being for people today, as well as leaving a world fit for future generations to make the changes they will deem necessary? Whether solutions come through introducing new technologies, changing consumption patterns or providing access to health care, water and sanitation, the fact is that any and all improvements depend on the co-operation of a number of different actors who interact in a complex and dynamic way. Moving from a traditional development m odel to one of sustainable development has been, and w ill continue to be, a transformation along these same lines. Whereas in the past most development decisions were driven primarily by economic considerations without regard to implications for the social or environmental spheres, the last 20 or so years of discussions around sustainability have transformed the way both public and private institutions conceive of growth, quality of life and other development-related concerns.

Citizens, Civil Society and Progress

Just as no inventor alone in the garage has the means to turn a discovery into a meaningful tool for society, no activist can alone achieve widespread social change. Each of them must communicate and interact with others to prove the merits of the new discovery or idea and convince others to adopt and promote it. Human advancement depends on an ongoing exchange between people and institutions. The decisions we make about how the world should be and how it can be improved depend on interactions among individual citizens, businesses, civil society and governments. These four categories function together in the complex and sometimes chaotic process of decision-making that we call politics.

The term civil society is one we hear a great deal today, one which like sustainable development can be hard to pin down to an exact definition on which everyone agrees. The London School of Economics Centre for Civil Society defines it as "the arena of uncoerced collective action around shared interests, purposes and values. The groups, associations and movements that make up civil society have played a part in all the important societal changes in the past century or more. Civil society organisations can be dedicated to specific issues or more general struggles. Indeed, they have been key to the success of very significant advances including universal suffrage, environmental protection, workers' rights and combating racial discrimination.

Sustainable development is no exception. Organisations such as the Sierra Club, founded in 1892 in the US, or Australia's Gould Group, dating from 1909, were advocating what w e w ould n ow term sustainability long before politicians and the m edia gave the matter much thought. Civil society organisations have been present at all of the major meetings that put sustainable development on the map. In fact, they have been instrumental in developing sustainability as an idea and in its translation into concrete practices. They have consultative status at UN and OECD meetings (inclusion that they had to organise and work for) and participate in policy debates. They do research, write policy briefs, and organise collective social action like protests and boycotts. They raise awareness and help educate the public and policy makers.

The Marine Stewardship Council (MSC) is a concrete example of what a civil society organisation can achieve in a domain like this. The MSC is an independent, global, non-profit organisation set up to find a solution to the problem of overfish ing. It was first established by Unilever, the world's largest buyer of seafood, and WWF in 1997. In 1999, it became fully independent from both organisations. The MSC works with fish eries, retailers and others to identify, certify, and promote responsible, environmentally appropriate, socially beneficial and economically viable fish ing practices around the world.

The MSC Principles and Criteria for Sustainable Fishing is an internationally recognised set of principles to assess whether a fish ery is well managed and sustainable. Only products from fish eries assessed by independent certifiers as meeting the standard are able to use the MSC logo on their products. For the first time, this gives consumers a way to identify – and the choice to purchase – fish and other seafood from well-managed sources.

What is the Role of Government?

A 2003 poll of Canadians showed that car salespeople are trusted less than almost any other profession, with only 10% of respondents finding them trustworthy. *Almost* any other: "national politicians" do even worse, at only 9%. Other people show similar opinions to the Canadians. Government itself is often criticised for a long list of failures, real or perceived: stifling innovation and entrepreneurship through taxes and red tape, caving in to pressure from lobbies and non-representative interest groups, leaving education or health care systems in poor condition. Governing in such a complex world is a huge challenge. Yet democratic governments at least try to make policies that will satisfy people and take care of important issues.

Before looking in more detail at the various tools governments can use, it's useful to recall what tasks governments perform in working for sustainable development. In general, through their data gathering and analysis, policy making and co-ordination, governments can provide support and leadership for moving society in a given direction. They can make sure that individual interests do not detract from the common good. Sustainable development contributes to this good, but actions to promote it may negatively affect the immediate interests of certain people, such as the shareholders of a factory that has to pay higher wages or install air and water filters.

Governments also intervene to deal with what economists call "market failures", situations in which market forces alone do not produce the most efficient outcome. The "externalities" mentioned on production and consumption would be an example of this – situations where the actions of one individual or group have costly consequences for others. Given the global nature of many of the challenges facing sustainability, nations have to co-operate at the highest levels to design and apply solutions. National governments have the authority and power to do this. They also have the means to ensure that decisions are applied. The three most important means by which governments can influence sustainable development (for better or worse) are regulation, taxation and spending. Each can play a role, but taxes tend to be more cost-effective and flexible than regulations, while subsidies are expensive for taxpayers and consumers.

Regulation

Governments may introduce new regulations in response to social or other pressures, but regulation can in turn have a marked effect on behaviour. Smoking in public places would probably continue without government intervention to ban it, for example. Good regulation is an essential tool for making sustainable development a reality. Social and economic conditions evolve, new materials and technologies are developed, and our understanding of health and environmental effects improves. We have to adapt regulations to correspond to changing conditions, and there will always be a need for new regulation. Nanotechnologies and biotechnologies hold great promise, but they also raise a number of questions regarding their safety and in some cases the ethical implications of their adoption. Governments have to gather and analyse the evidence and see if there is a need to change or create regulation. Their decisions will have a major impact on how these technologies and the industries that use them develop.

Nanotech and biotech reveal one of the weaknesses of regulation – the pace of change in some areas is far faster than the pace at w hich the regulator works. In other cases, governments may try to move more quickly than the electorate is prepared for – many people are hostile to changes in legislation that affect their working conditions or pensions for instance. Regulation has other limits, too. If, for example, bans were 100% effective, there would be no illegal drug use, no speeding, in fact no crime or delinquency to worry about at all. Moreover, the way bans, restrictions, standards and other types of regulation are drafted and applied can also cause problems, leading to counterproductive "red tape". Instead of providing a coherent framework for activity, red tape hinders innovation, stifles initiative and adds unnecessary administrative burdens to economic and social activity.

Regulation can however lead to desirable outcomes for sustainability and increase individual well-being and that of society as a whole. Vaccination and other public health initiatives are good examples, as is the obligation to educate children. We take some of these regulations so much for granted that we may be surprised to learn that they are comparatively recent and had to be fought for, such as regulations concerning the quality of drinking water and food or the safety and environmental impact of automobiles.

Regulation then is not inherently good or bad. This point is recognised in a set of guiding principles for regulatory quality and performance established by the OECD. The importance attached to identifying how any proposed changes to regulation might affect other policy objectives is especially important for sustainable development, where changes in one area may have important consequences elsewhere. The principles also stress that regulation affects, and is affected, by other types of intervention, notably government spending and subsidies and taxes.

Spending

Governments are big spenders and the way they allocate funds influences practically every aspect of the economy and society. This can have direct impacts on sustainability. A government with a certain sum to devote to transport can decide to invest it in improving the road network or in developing rail services. It can use the energy budget to build new electricity generating capacity or to promote insulation and other energy-saving technologies. Health spending can focus on developing innovative therapies or on preventing common pathologies. International aid can be used to encourage bilateral trade or to promote technical co-operation. In everything from science budgets to welfare programmes, the choices governments make have an impact.

Many OECD governments subsidise fossil energy, and removing or reforming these subsidies would help policies to tackle climate change. Agriculture may seem a less obvious example, but it is one of the main beneficiaries of subsidies. Consumers and taxpayers transfer over $300 billion to OECD agriculture each year. Some of this is used to help improve agricultural techniques or quality, but much of it keeps prices high. For example, despite reforms, average OECD domestic prices for rice, sugar and milk are still more than double those on world markets, which is particularly hard on poorer consumers who spend proportionately more than the rich on food.

Historically, the goal of farm subsidies has been to increase production and therefore food security for a given nation. Over the course of the 20th century, this has meant increasingly mechanised agriculture, a shift towards single crop (monocrop) cultivation, heavy reliance on fertilisers and pesticides, and depending on climate, drainage and irrigation schemes. This so-called "high input" agriculture resulted in a boom in production. At the start of the 20th century, an American farmer had to feed on average 2.5 people in the country. Today, a farmer feeds over 130 people according to the National Academy of Engineering, and estimates that include exports are even higher.

These advances have major impacts on the environment and on farming communities:

- Highly mechanised agriculture can result in increased soil erosion, as machines break up the soil. This results not only in a loss of fertility locally, but also in water pollution as these sediments run off the surface.
- % Conversion from small, diverse fields with hedges to monocrop reduces the niches available to insects and birds. European farmland bird populations have declined by 40% in the last 30 years, and for all but a few species that trend is continuing.
- % Nutrient pollution (eutrophication) is the leading water pollution issue. In most areas, farms are the largest source of the nitrogen and phosphorus at the root of these harmful algal blooms.
- % Previously pure sources of groundwater are now contaminated by pesticides that have leached through the soil from farms above.
- Irrigation is the largest human use of freshwater, accounting for over 70% of the total worldwide. Reduced river flow and dropping groundwater levels make this use a potential source of conflict. In the case of rivers and other surface waters, habitat for fish and birds is sacrificed to maintain food production.

"Decoupling" aid from production is a key measure: goals for agriculture are changing, and subsidies can be a powerful tool for reaching those new goals. Once again, the food crisis illustrates how numerous strands are interwoven. High prices weaken the case for subsidies and could enable funds to be freed for other uses. But high prices also encourage farmers to produce more. They may as a result abandon schemes to leave land uncultivated so that it can be used for other purposes, such as to encourage biodiversity. Carefully targeted subsidies can help to restore the balance among various policy objectives: this requires transparency regarding who benefits and who pays for subsidies such as the European Union's Common Agricultural Policy (CAP), and careful co-ordination between the many stakeholders.

The impacts of agricultural subsidies (positive or negative) obviously touch the social sphere as well as the economy and the environment. Indeed, it is hoped that the reform of agricultural subsidies will allow farmers from developing countries to compete in the global market. This potential for profit should encourage the development of farming infrastructure in countries that have not traditionally exported, with important implications for local employment, purchasing power and food supply. Food security is once again a concern – worldwide – and all governments will have to develop appropriate measures for encouraging productive and sustainable agriculture.

Taxation and Emissions Trading

The flip side of the spending coin is, of course, taxes. When we think of taxes and sustainability, so-called "green" or "ecotaxes" come to mind first, since these (like emissions trading) are designed to contribute directly to environmental sustainability by making "bad" environmental behaviours more costly. However, as we've argued throughout, the environment is only one part of the process. The social and economic aspects of sustainability are influenced by taxes, too, and in fact are among the biggest items in national budgets. Education for example represents 5% of government spending in OECD countries on average, while health accounts for another 6%. But since "social taxes" existed long before the concept of sustainable development was invented, and their role is rarely presented in this light, their importance is easy to overlook. Nonetheless, through mechanisms such as social welfare schemes, they play an essential role in addressing issues that market mechanisms and private initiatives alone cannot deal with efficiently.

Likewise, taxes are often perceived as hindering economic development, but governments use them and the revenues derived from them to shape and promote economic development. The social and economic roles of taxation overlap in many cases too, as when funds are invested in developing certain sectors or regions, or when social measures are used to ease or encourage the transition from traditional to new activities.

Interestingly, for many sustainable development issues there is a very strong argument to be made in favour of using taxation and other market-based mechanisms *instead* of subsidies: what are the chances that policy makers will identify every initiative worthy of support and make the appropriate subsidy, without accidentally supporting some initiatives which turn out to have negative effects? On the other hand, a very simple taxation mechanism can spur innovation on the part of businesses, as they come up with their own solutions to reduce a particular practice.

There are several reasons to use economic tools for sustainable development:

- They can provide incentives for behaviour that fii ts with sustainable development goals and deter actions that go against those goals.
- Overall environmental, social and economic costs could be built into prices using such measures, driving markets towards a more sustainable economy. They encourage innovation by providing market pressure.
- The revenue generated could be used to reduce other taxes or finance social measures.

A May 2008 Chicago Tribune article put it like this: "They [consumers] can pay high prices to oil producers or to themselves. The tax proceeds can be used to finance programs of value here at home or to pay for cuts in other taxes even as they curb the release of carbon dioxide."

Putting Sustainable Development to Workin Governments

The governments that signed Agenda 21 at the Rio Earth Summit expressed a certain degree of optimism about sustainable development. For them, the role of government would be central in achieving those goals. It makes sense: sustainable development is a concept with the potential to change many things for the better, but if not firmly anchored in policy-making bodies at all levels of government local, regional, national and international concrete achievements will remain elusive.

In the same way, if policies within one government ministry undermine those in another, progress stalls. Before promoting large-scale tourism for example, it may be wise to ask if the golf courses and swimming pools will mean there's no water left for farmers. On the other hand, if you favour agriculture over tourism, you may lose the chance to create hundreds of jobs in an area with high unemployment. Governing for sustainable development doesn't mean favouring one aspect and neglecting the others; it's about finding the most coherent balance among different claims and devising the most efficient administrative and other means to implement strategies.

But how do you go about making plans for what you would like to accomplish? Agenda 21 signatories agreed to develop National Sustainable Development Strategies (NSDS), documents intended to fit the specific needs and goals of different countries while addressing the basic sustainable development priorities that the international community (OECD and UN) has agreed on. Given the flexibility allowed, strategies vary widely. Most OECD countries now have an NSDS in place, each with particular strengths and weaknesses. So over fifteen years on from Rio, how are they doing? Are certain countries or regions leading the way? If so, how do they do it?

A recent OECD workshop on best practices for institutionalising sustainable development gave some concrete suggestions. Participants identified a number of indicators of success such as inscribing sustainable development in constitutions and legislation and including it in national budgeting processes.

An essential part of a programme's success is its perceived importance. For sustainable development to be taken seriously, it needs to be centrally located in a ministry or department with influence across all government activities in the Prime Ministry as in the case of Austria or in the Ministry of Finance as in Norway. When sustainable development is "anchored" in one of these central functions, its impact is enhanced and more easily coordinated throughout the different parts of government. Sustainable development can also have its own ministry as in the case of France.

In the case of new sustainable development ministries, a diverse range of concerns previously separated across government ministries are re-grouped into one. Putting energy, ecology, maritime affairs, territorial planning, forestry and other domains together allows for integrated analysis and decision-making and makes it easier to avoid the pitfall of policies that contradict and undermine each other. Yet this approach can only be effective if supported by the prime minister or presiden in other words, if its recommendations translate into concrete implementations.

New Zealand shows how the social dimension can be included. The Sustainable Development in New Zealand programme gives equal weight to social sustainable development in relation to the economy and environment, with special attention to demographic trends, new roles of women in society, improvements in health and housing, and better integration of Maori communities.

What does Governance for Sustainability look Like?

Intergenerational questions are an important component of the social dimension, which is why the *Swedish Strategy for Sustainable Development* adopted an intergenerational timeframe which includes a vision for the future which should remain valid for a generation or at least 25 years.

Sustainability in all Levels of Government

Leadership at the national level is one key part of governance for sustainable development. However, initiatives a t regional and l ocal levels are also critical to its success. After all, local governments have the closest proximity to what people and businesses actually do how they pollute, how they produce and consume, how they experience health care and education systems. People usually decide to take action on a g iven issue because of what they perceive in their immediate environment and local governments have a lot to do with how a place looks, feels and functions.

Local governments have to identify the critical relations among many factors likely to shape economic, social, political and environmental quality. But even the city level administration may be too remote from the day to day impacts of decisions. Effective governance also needs lower level local networks that include non-governmental actors, associations and businesses, for example to deal with social tensions or make the most of economic opportunities. As the UK Commission for Sustainable Development says: "National policy sets direction, but it's practical action at the local level that makes sustainable development real."

Identifying the correct level of government for addressing a question is itself an important and often complex task. Large cities or metropolitan regions, for example, regroup a number of localities with divergent views on issues important to the greater metro area, as well as different ways of dealing with the range of problems cities handle. Also, many sustainability issues are "regional" in nature think of air pollution or land use. Coherent governance for sustainable development for these large urban areas often requires a regional institution that can co-ordinate efforts and solve inconsistencies in local policies and initiatives.

Furthermore, strategies that are seen as simply one more government programme imposed from above have less chance of succeeding than those defined through consultation and debate.

It would be unrealistic to imagine that everybody would be satisfied with every aspect of a national strategy, but the strategy is more likely to be implemented if everyone concerned has a chance to influence outcomes. This is why the Czech Government Council for Sustainable Development includes government, business, academics, NGOs and other stakeholders and serves as the umbrella group for developing, implementing and revising the national sustainable development strategy.

Governance of Uncertainty

The media often emphasise the role of corporations and individuals in sustainable development – after all we're the ones building, purchasing, and so on – but governments play an equally significant role and can have far more influence than even the biggest multinational. Their ability to influence behaviours and co-ordinate efforts can make all the difference in producing substantial results. If not coherent, though, government actions can be a barrier to improvement.

When describing the role of government, it's easy to give the impression that governance for sustainable development is merely a matter of identifying objectives then putting in place a series of measures and bodies to oversee them. It's not. Just about every aspect of the economy, society, and the physical resources on which they ultimately depend, influences sustainability.

Outcomes depend on an infinite number of interactions working on different timescales of varying importance. No model, however robust, no foresight, however penetrating, can tell us everything we'd like to know. Governments attempting to implement sustainability have to deal with this uncertainty. Not only their goals, but the strategies and instruments used to achieve them must be sustainable, too. They must be rigorous enough to be effective, but flexible enough to adapt as circumstances and priorities evolve. In the face of uncertainty, governance itself has to be sustainable.

References

Maddison, A. (2001), *The World Economy: A Millenial Perspective*, OECD Publishing, Paris.

OECD (2003), *Emerging Risks in the 21st Century: An Agenda for Action*, OECD Publishing, Paris.

Gurría, A. (2007), "The Economics of Climate Change: The Fierce Urgency of Now", speech at the UN Climate Change Conference, Bali, 12 December 2007.

IEA (2008), *Energy Technology Perspectives 2008: Scenarios and Strategies to 2050*, OECD Publishing, Paris

6

Governing Climate Change

Contemporary environmental risks, from radio-active waste to genetic modification, air pollution to soil degradation, attract the sustained attention of geographers interested in how such issues can be managed and mitigated, how they are socially constructed and the political conflicts they provoke. Developments within social theory, in particular related to the work of Ulrich Beck, have added food for thought to these geographical concerns.

Significant changes in global average temperatures and regional climatic conditions are to be expected, and that there are possibilities for unpredictable alterations to the climate system. The issue of how to address the risks posed by climate change has been another area of contention, and has spawned international negotiations, domestic strategies and public concern throughout the 1990s. The challenges of governing climate change have been apparent as nation-states struggle to come to international agreement and take domestic action. These struggles have been particularly evident in contexts where environmental and economic interests are seen to be in conflict, such as the USA and Australia, as the recent withdrawal of support by the Bush administration for the Kyoto Protocol illustrates all too clearly.

Risk Society

Beck's thesis concerning 'risk society' ranges far and wide: from the workplace, to family, class and community, and on to scientific laboratories, social movements and the corridors of power. At the heart of his explanation concerning the dynamic of change from industrial modernity to an emergent risk society are the new, primarily environmental, risks created as the unintended consequence of enlightenment rationality. In placing rationality at the centre of his thesis Beck draws on the heritage of critical theory and its concerns with the dialectics of the enlightenment, the role of science and technology in modernity, and the need to create a renewed civil society.

Risks

Several features of the issue of climate change epitomize Beck's definition of the experience of risk in risk society (. First, the causes of climate change lie deep within modernity. Apparently innocuous and invisible gases, such as methane and carbon dioxide, released as by-products of development (the industrialization of agriculture, increased energy consumption and economic growth) change the composition of theatmosphere with untold consequences. This accords with Beck's supposition that in risk society, risks arise not from a lack of modernity, as hazards associated with poverty and underdevelopment might be conceived, but rather as the side-effects of modernization. In his later work, Beck argues that the 'world risk society' is beset by different categories of risk, those created through development, those which arise from a lack of development, and those which ensue from conflict situations. Despite this acknowledgement, it is the appearance of risks as 'unintended consequences of modernity' created by, and providing the dynamic for, reflexive modernization that forms the core of his thesis. Second, the sources and experiences of climate change risk are indeterminately distanciated over space and time, stretching social and natural relations of cause, effect and responsibility, and transcending 'the spatial, social and temporal limits of industrial risks of the past'. While developed countries are responsible for the vast amount of emissions of greenhouse gases to date, the impacts of climate change will be felt across the globe and by future generations. However, as Beck suggests, such risks are inescapable; while the impacts of climate change will be felt most in societies which are more vulnerable, the 'boomerang effect' means that developed countries will also experience risk, through changing climate patterns and their associated impacts.

The third dimension of climate change risks is intimately connected to the second; because of the distanciation of causes and effects risks are based on 'causal interpretations, and thus initially only exist in terms of the (scientific or anti-scientific) knowledge about them'. Although climate change may be sensed by individuals, our understanding of the processes through which greenhouse gases affect climate systems, the modelling of future changes, and the monitoring and recording of global climate trends is dependent on scientific understanding. This reliance on scientific mediation of risk, Eden explains, leads to the 'scientization of environmental problems: they are identified and primarily constructed through the application of scientific techniques and reasoning'. Simultaneously, the reciprocal process occurs so that 'science becomes politicized and drawn into policy formulation'. As science, in the form of the IPCC, was drawn into the policy arena concerning potential courses of action about climate change, the demonopoliz-ation of expertise occurred and debates over whether the evidence warranted action dominated policy arenas. This accords with Beck's proposition that, given the nature of contemporary risks, 'political conflicts are increasingly turning on the possession and articulation of knowledge'. Given these characteristics, climate change can be taken as a paradigmatic example of the kinds of contemporary environmental risks Beck argues form the focus of risk society. Whether these risks have engendered new social and political processes characteristic of risk society therefore merits further investigation.

Reflexivity and Reflection

The transition from 'industrial society' to 'risk society' Beck describes as reflexivity: modernity undergoing a process of (unseen) 'self-confrontation'. Risks are produced within modernity, but their distanciated, unpredictable and imperceptible nature means that the institutions of modernity can not 'comprehend or legitimize' them. The politics of risk society shifts from concern with the distribution of 'goods' to issues surrounding the distribution of 'bads'. This in turn renders class and the relations of production less salient as the basis for 'political interest formation and organization' than the 'relations of definition' through which risk is understood and delimited. The dynamic of risk society in Beck's view is underpinned by a division between the 'political realm', the traditional (state) institutions of accountability and control, and the 'subpolitical realm', conceived as politics 'outside and beyond the representative institutions of the political systems of nation-states' where previously 'private' decisions are rendered 'political'. In risk society the political institutions of modernity are irrelevant, inadequate or impotent as decision-making power, control and legitimacy increasingly locate outside the political system in economic, technological, scientific, community and consumption 'sites which were previously considered unpolitical'.

Initially the institutional crisis which these processes engender goes unnoticed. However, within the process of reflexivity Beck proposes lies the possibility for greater social reflection leading to a second stage of risk society. While society 'still makes decisions and acts on the pattern of the old industrial society... debates and conflicts which originate in the dynamic of risk society are.. superimposed on interest organizations, the legal system and politics'. The institutions of modernity are unable to remain stable in the face of new 'conflicts of accountability' as disputes erupt over how the consequences of risk can be 'distributed, averted, controlled and legitimated'. Beck offers three scenarios of how the second stage of risk society might take shape. First, and currently dominant, that the division between 'political' and 'subpolitical' realms is further entrenched ensuring the continuation of industrial society. Second, that technological and economic development are democratized through conventional political systems which fail to account for subpolitics and therefore also lead to the continuation of industrial society. Third, and the focus of Beck's later work, that the emancipatory potential of risk society is realized through a differential subpolitics, supported by innovative legal and communicative institutions. However, there is a fair degree of confusion and ambiguity in Beck's writing over the political consequences of reflexive modernization and the evolving 'subpolitics' of risk society. On the one hand, it is suggested that a 'division of labour' occurs so that 'the truly political disappears in and from the political system and reappears... as subpolitics' leaving formal political systems and institutions unable to fulfil their roles of decision-making, control and accountability. On the other hand, Beck argues that although the 'unbinding' of politics creates subpolitics, the institutions of the political system retain power and influence and are thrown into coalitions with subpolitical institutions and actors. From this morass of evolving ideas, three key points concerning the political possibilities of contemporary risk emerge: first, that the politics of risk society are not conducted only, or primarily, through the formal political system; second, that in the light of risk society, the formal political system is weakened; third, that it is within subpolitics that conflicts of accountability will be resolved or ignored. Whether, and with what effect, the politics of climate change risk conform to these characteristics is considered in detail below.

Critics have taken issue with much of Beck's thesis concerning the nature of risk and society. Though it is not my intention to deal with these comprehensively, several points are pertinent to this discussion. The risk society has been seen as borne out of the particular circumstances of postwar Germany, and its appropriateness for describing the politics of nature and society outside western Europe, or even Germany, questionable. While risk society seems to describe the nature of climate risk, whether other environmental risks are as distanciated and imperceptible is debatable. Moreover, although the risks of industrial modernity may well have 'assaulted... the senses', this did not and does not preclude their mediation through expert and public understandings. This has led some commentators to suggest that there is little novel in the contemporary challenges posed by risk to political and expert communities that Beck describes. Bennett argues that risks have always been socially constructed, and that the risks of industrial society 'appear certain today because the disputes over their insurability have... been resolved' rather than because of any difference in the intrinsic quality of the risks themselves. Gandy also supports the view that throughout the modern period 'ecological risk and public health scares have been repeatedly translated into new regulatory regimes' which question the status quo, though he argues that the nature of contemporary risks do pose distinctive challenges. Any temporal distinction between an 'industrial society' and a 'risk society is therefore perhaps unwarranted. Instead, it is possible to argue that (environmental) risk has continually created institutional crises, and that some contemporary risks pose particular challenges for the institutions which structure modernity. A further problem with the concept of risk society concerns its underlying dynamic: reflexive modernization. Critics have argued that the dialectic between reflexivity and reflection is not fully explored and the emergence of risk society in particular contexts needs further explanation. The applicability of risk society to all cases of contemporary risk politics is not universal; rather, the usefulness of Beck's insights into the dynamics of risk need to be demonstrated in each case. The challenges posed by climate change do indeed appear distinctive. The unbounded nature of climate risk has profound implications for society, rendering environmental regulation based on national borders and future predictability impotent, and leading to new conflicts which stretch social relations over space and time. In drawing attention to the importance of the processes through which risks are defined and delimited, Beck's work challenges those accounts of the politics of global environmental change which take interests as given. Defining environmental risk, and its ensuing obligations, is a means through which a range of scientific, social and moral concerns are articulated , and through which social structures and institutions are challenged and reproduced. The following sections consider how climate change obligations and responsibilities were contested and constructed within the Australian policy process in order to assess how the politics of risk take shape.

International Negotiations

The interpretation of intergenerational justice with respect to climate change is both interesting and complex, involving concerns for the representation of 'distant strangers' within contemporary environmental politics and raising difficult philosophical issues about the needs, deserts and identity of future generations. However, as Grubb argues, in practice the interpretation of intergenerational justice has provided a simplified rationale for action; the need to 'prevent dangerous anthropogenic interference with the climate system', or, in other words, to avoid future harm.

In contrast the debate over international equity, involving the consideration of climate impacts, and how they can be averted or compensated for, and sources of emissions, and how they can be mitigated, has been heated. While the impacts of climate change remain uncertain, it is clear that some places will experience more risk than others, notably the small island states and delta regions, and debate has raged over how such impacts can be compensated. The UNFCCC states that developed countries should assist developing countries with the task of adaptation, a premise which has been reaffirmed in the Kyoto Protocol , though there is considerable ambiguity as to what this will mean in practice. By far the most long running and divisive debate concerning equity and climate change internationally has been the respective responsibilities of nation-states for reducing emissions of greenhouse gases. In the main, this debate has been cast in north-south terms, with the argument made that the north should take action (first) because of its responsibility for the vast majority of emissions to date, its high levels of per capita emissions and its capacity to take action . The UNFCCC recognizes these principles, and also states that technical and financial transfers from the north are necessary to enable emissions reductions to take place in the 'south'. The inclusion of such principles goes some way to countering claims that the politics of climate change represents a form of 'environmental colonialism' , though it does little to finally resolve the issue of respective responsibilities. Throughout the Conference of the Parties (COP) process, the USA, Australia and several industrial lobby groups argued that developing countries should be included in emissions reduction commitments because of their growing contribution to the problem, concerns about international competitiveness and environmental effectiveness, as well as attempts to shift blame from themselves for the slow progress of negotiations. While this proposal was not explicitly incorporated into the Kyoto Protocol, more effort has been directed towards thinking of the ways in which developing countries could be included in future agreements. The scope and structure of these debates suggest that the international politics of climate change risk have not been conducted only in terms of material interests, which in any case are difficult to identify, but on the basis of the kinds of conflicts of accountability for risk and its mitigation that Beck argues dominate risk society. Such conflicts have also been visible in debates concerning the role of individual countries in addressing climate change. The distribution of responsibilities between countries of the 'north' became the subject of sustained debate in the run up to Kyoto. Within the UNFCCC it is repeatedly stated that obligations can fall unevenly on different parties. The text refers to 'common but differentiated responsibilities and respective capabilities' , and the need to take into account 'differences in these Parties' starting points and approaches, economic structures and resource bases, the need to maintain strong and sustainable economic growth, available technologies and other individual circumstances'. While such reservations were included to indicate that the 'developed country Parties should take the lead in combating climate change' , they have been used to good effect by those arguing that the impact of addressing climate change will not be felt evenly amongst the countries of the 'north'. To this end, a key article in the Convention, referred to as the 'fossil fuel clause', urges that Parties:

> Take into consideration in the implementation of the commitments of the Convention the situation of Parties, particularly developing country Parties, with economies that are vulnerable to the adverse effects of the implementation of measures to respond to climate change. This applies notably to Parties with economies that are highly dependent on income

> generated from the production, processing and export, and/or consumption of fossil fuels and associated energy-intensive products and/or the use of fossil fuels for which such Parties have severe difficulties in switching to alternatives.

This clause provided Australia, amongst others, with significant fuel for their argument that international obligations should be differentiated on the basis of the predicted impact on the economy of taking action. Rather than the transition proposed by Beck from the 'relations of production' to the 'relations of definition' as the basis for politics in risk society, the international politics of climate change risk illustrate how definitions of obligation and responsibility are shaped by discourses of economic costs and benefits alongside moral and scientific considerations.

In the initial stages of negotiating climate change responsibilities the Australian Federal Government was keen to show its willingness to undertake global environmental obligations, and adopted a national target for emissions reductions and the *National Greenhouse Response Strategy*. As the introduction to this strategy makes clear, these obligations stem both from formal commitments under the UNFCCC and from the potential risk of climate change to 'Australia's natural, social and working environment, as well as... [to] the global community and global environments'. However, as illustrated in the statement above, from their inception climate change responsibilities have also been seen to involve the protection of Australia's economic interests, defined primarily as those concerning the energy and energy-intensive industries. As the analysis below illustrates, these concerns have structured the identification and assessment of risks and have controlled the 'relations of definition' through which the obligations of climate change have been understood.

Having initially backed Australia's international commitments under the UNFCCC, resource and energy industries began to urge the Federal Government to take a more cautious stance as the climate negotiations entered their second phase. At COP-1 in Berlin, Australia joined with the USA in opposing the need to strengthen the commitments of the UNFCCC without further participation by developing countries. The successful counter argument put forward by the EU, with the support of the majority of developing countries, was that legally binding targets for emissions reductions should be agreed by COP-3 at Kyoto in 1997. The acceptance of this position by the Clinton administration at COP-2 left Australia isolated internationally with an increasingly entrenched view that any international obligations should be tempered by recognition of scientific uncertainty and the potential economic costs of action:

> I went to COP-2 as a representative of an Australian NGO.... It was rather like being a temperance campaigner on a pub-crawl. The Australian delegation was committed to, and in most cases quite comfortable with, the government line that: a) there is not a serious problem; and b) if there is Australia shouldn't have to do anything about it because of our economic structure and the fact that our population is so spread out.

To counter this position, environmental groups, scientists and some sub-national governments argued that Australia should shoulder its responsibilities squarely in the international arena, stressing the potential consequences for Australian health, welfare and environments should climate change continue unchecked. This plea had little impact and the Federal Government continued to argue that in defining its climate obligations with respect to domestic economic concerns it did not deny responsibilities, merely ask that they be fairly attributed according to the Convention's own provision for 'common but differentiated' obligations. Through insistence of Australia's economic vulnerability and small overall contribution to global emissions, the Federal Government repeatedly stated that it 'would not 'sell out' the national interest' to take on undue responsibilities. In taking this stance the Federal Government is in effect aligning the interests of 'the nation' with the concerns of the resource and energy industries that taking action on climate change would be more economically damaging for Australia than other OECD countries:

> Australia is not seeking to be exempted from the Convention. It is seeking a fair go – that is an outcome where the onus of response measures falls no more heavily on Australia than on other countries. Without such an outcome it is not realistic to expect agreement to be achievable and sustainable.

The justification for these arguments lies in economic modelling which suggest that Australia would suffer significant job losses, reduction of GDP and savings losses in the order of A$9000 per Australian in complying with a uniform target for emissions reductions. These findings have been contested, in terms of their ability to account for the potential economic savings of energy efficiency measures and the environmental costs of inaction, and their objectivity questioned due to the role of resource industries in sponsoring the modelling project. Despite this, they have been used by the Federal Government as the basis for arguing that an equitable division of climate change responsibilities would have to allow 'differentiated obligations' on the part of developed countries in light of their economic circumstances. The concept of differentiation was touted not only as creating the possibilities for 'fair and equitable' outcomes, but also outcomes that would be environmentally effective by ensuring participation in, and compliance with, the resulting Protocol:

> Australia is heeding the message of the scientists.. We accept responsibility for our share of the greenhouse task, but we believe that a desirable environmental outcome can only result from policies which are achievable, effective and equitable.

Meeting obligations for the global environment is viewed as an economic exercise. In turn, existing relations of production shape the ways in which climate change obligations are defined. The absence of any *environmental* designation of responsibilities was noted not only by environmental groups:

> Australia is campaigning to be allowed to substantially increase its emissions while the rest of the developed world reduces them. With Australia one of the highest emitters on a per capita basis, this flies in the face of the principle that those who cause a pollution problem should take responsibility for cleaning it up.

but also by the *Australian Financial Review*, a paper not known for its green leanings:

> Suggesting that all countries should bear an 'equitable burden' in reducing global greenhouse emissions conflicts with the efficiency based 'polluter pays principle'. It is... stretching the issue to argue that lucky resource-rich – and hence emissions-intensive – economies should be given a special deal because their abatement costs are relatively high

During 1996–97 the principle of differentiation received widespread comment, criticism and varying levels of support within the international arena and Australia. Some major players in the climate change negotiations, notably the EU and the USA, initially saw the idea of differentiation as internationally untenable. However, the EU advocacy of differentiation within the EU 'bubble' made its opposition to differentiation internationally increasingly difficult to sustain. Pragmatically, differentiation was also thought too complex to use as a negotiating principle, with disputes arising over whether it should be the energy-efficient nations, such as Japan and Norway, or particularly fossil-fuel-dependent economies, like Australia, who should bear fewer emission reduction responsibilities. In the first half of 1997 the Prime Minister, John Howard, endeavoured to 'push the principle of differentiation on every available occasion before every world leader in every forum' that he could find . Despite this marathon effort, there were significant fears expressed by some in the resource and energy sector that without international acceptance of differentiation, Australia would have to reconsider its continued participation in the climate negotiations.

In late 1997 this predicament receded. The withdrawal of US support for greater emission reductions in October re-opened the question of exactly what climate change obligations entailed. In November, the Federal Government indicated that, under a new strategy for action, emissions were expected to increase by 18 per cent over the period 1990–2010, instead of the 28 per cent predicted. Controversially, the Federal Government entered the Kyoto negotiations with the expectation that their obligations for protecting the global environment would allow them to increase emissions of greenhouse gases. In the days immediately prior to Kyoto 'the possibility of limited, carefully bounded differentiation' attracted support from the USA, Japan and others looking for a way round the proposed uniform target of a reduction of 15 per cent by 2010 put on the table by the EU. The negotiated Kyoto Protocol commits developed country signatories to an overall reduction target of 5 per cent below 1990 levels in the period 2008–12, through the pursuit of differentiated targets. However, the fact that differentiation was adopted owes more to political expediency than any recognition of the 'principled position' Australia had taken. The exact numbers assigned are the result of bargaining amongst nation-states over individual targets in conjunction with other parts of the Protocol, such as the inclusion of emissions trading, the comprehensive approach taken to all greenhouse gases and the incorporation of sinks. In this melee, Australia maintained its case for 'special treatment', and was one of the only three countries assigned the responsibility of containing emissions increases, at 8 per cent above 1990 levels.

In addition to differentiation, a key tenet of the Australian approach to designating international obligations throughout the COP process has been that developing countries are included in the allocation of emission reduction responsibilities. While the UNFCCC makes it clear that developed countries should 'take the lead' because of their historical responsibility for emissions and capacity to address the problem, the resource and energy sector, alongside parts of the Federal Government bureaucracy, questioned this approach. Firstly, in the light of the future contribution of developing countries:

> If you really believe that the greenhouse problem is a problem, then you have to... find some policy that is going to allow you to engage developing countries. Because... the lion's share of all of the increase in emissions is coming from developing countries... so to be practical about it, you've got to have some means of getting [developing] countries in (Australian Bureau of Agriculture and Resource Economics.

and secondly on the grounds of containing 'carbon leakage', as multinationals place their operations outside the jurisdictions of any Protocol:

> If you went to a company... involved in resource development and you said to them, look, what would happen to your investment decisions if Australia was forced to comply with stringent targets? They... would tend to say... Australia will just be not attractive to us and so we will invest in other countries. Then you have to say what is going to happen to energy consumption and greenhouse emissions if that happens? And depending on where you go and what sort of energy source you use of course, the effect on greenhouse may well be negative... if you went to India or China for example where you're using again a lot of coal, perhaps with poorer technology.

Far from heralding a fragmentation of politics, the risks of climate change have provided an additional arena through which the inter-dependencies between government and the 'subpolitical' resource and energy sector are intertwined. Arguments about the future role of developing countries and the need to prevent carbon leakage, despite questions raised over their validity, were accepted, and indeed reinforced, within the Federal Government. In turn, they continued to push for the inclusion of developing countries in the Protocol, either through the development of 'joint implementation' or more formal obligations, using similar justifications:

> If we were to stabilize emissions in the energy sector by 2010 without any complementary obligations in developing states, up to [A]$12 billion worth of planned investments in these sectors could move offshore. The Government simply cannot ignore the human cost these losses would impose on Australian families.

In this division of responsibilities the Australian Federal Government again finds itself sharing common ground with the US Senate, which voted unanimously that the USA should not accept any Protocol that did not include commitments from the developing countries. The Kyoto Protocol steered a path between opposing views through a continued emphasis on technology and financial transfers and the use of the 'clean development mechanism'. In the three years since the creation of the Kyoto Protocol, the respective responsibilities within and between 'north' and 'south' have continued to dominate the debate as key elements of the Protocol, such as the use of carbon sinks, the clean development mechanism and carbon trading, remain unresolved. The fragility of the balance reached at Kyoto has been all too apparent with the recent withdrawal of support by the Bush administration.

International negotiations and institutions have provided an arena in which nation-states, sub-national and supra-national organizations 'inter-subjectively come to some understandings about what norms concerning global warming mean'. Through these processes conflicts of accountability have surfaced as liability, blame, responsibility and rights have been constructed and contested and the risks that climate change poses stretch existing 'relations of definition' to their limits. The position taken in these debates and conflicts by the Australian Federal Government owes much to the dominant understanding of responsibilities – primarily to certain economic sectors and regions – held by the resource and energy industries, and some parts of the state at national and sub-national levels. In turn, these interest groups relied on the Federal Government to negotiate a favourable outcome through the international negotiations. The prevalence of economic considerations as a key discourse in the international negotiations during the late 1990s suggests that Beck's conceptualization of the dominant forms of knowledge and expertise in risk society is too narrow, and that the relations of production are central to the politics of risk.

Furthermore, a distinction between the 'political' and 'subpolitical' is difficult to draw. In this case, the social and political processes which arose in response to climate change risk served to deepen interdependencies within a coalition formed between the formal political system and subpolitical/ private economic interests, which engaged in the definition, control and legitimation of the concepts and practice of international responsibility. New relations of definition have been mooted within a wider policy network , for example by Greenpeace and the Australian Financial Review that the polluter pays principle should extend across international borders. However, these new discourses have failed to steer the definition of obligations away from the national interest, the economic survival of the resource industries and the economic prosperity of the present generation towards distant strangers, other industries, or a broader definition of well-being for the present generation. The choice of a spatio-temporal framework through which obligations and responsibilities are delimited is not neutral. Rather, it represents and embodies a particular set of values about both environment and society which privileges certain places, economic sectors, and people.

Climate Partnerships

The international conflicts of accountability documented above did not occur within a vacuum, but were influenced by domestic debates over the extent of environmental and economic risks, and how risks and responsibilities should be shared within Australia. The evolution of domestic climate change policy within Australia saw the development of a 'partnership' approach, summarized above by Senator Hill, as it became clear that the Federal Government was unable to address climate change obligations alone. As Beck's thesis would suggest, in the face of climate change risk, the state appears inadequate or even impotent. However, as the discussion below illustrates, this does not mean that the formal politics of the state gives way to the subpolitical governance of risk.

Despite an extensive consultation during the formation of domestic climate change policy, the final stage of securing policy commitments remained closed to participation from those outside federal bureaucracies, and the resulting *National Greenhouse Response Strategy* (NGRS) was primarily seen as a document produced by, and for, Federal Government. However, many of the measures contained within the NGRS, from energy-efficiency labelling of appliances, through investment in renewable energy technology, to decisions concerning land use and transport, rely more or less explicitly on other agents and institutions. It is clear that Beck's supposition that the state is dependent on the subpolitical sphere in order to address risk is applicable in this case. However, the nation-state, and other levels of formal political governance, are still central to this process. Although the Australian constitution gives the Federal Government powers to intervene on environmental matters in the context of trade or international obligations

, the primary decision-making power rests with state governments, with local governments playing a more minor role. The exacting task of implementing the NGRS was dependent on the co-operation of state governments, which was not forthcoming, and local governments, which were sidelined from the strategy. Support for the strategy amongst industry and the community was also negligible, with many measures ignored and few suggested regulations imposed. Given this context and a lack of coordination between jurisdictions and policy areas, it is hardly surprising that progress in effecting the NGRS was seen at best as uneven (National Greenhouse Advisory Panel and at worse as 'glacial'. As the NGRS moved off the drawing board and into action, conflicts of accountability over climate change responsibilities within Australia were endemic. Nowhere was this more marked than in the brief debate over the imposition of a carbon tax which surfaced in 1994. The resource and energy industries argued that this approach would be contradictory to the aims of the NGRS that the costs of taking action to reduce emissions of greenhouse gases would not impose an undue economic burden on particular regions or industrial sectors. In response, they sought to redefine their responsibilities by arguing that a voluntary partnership approach was the most appropriate course of action given the state of scientific knowledge and potential costs of regulation. In the wake of this debate, and the failure of the NGRS to progress climate change policy, the Federal Government launched the additional *Greenhouse 21C* strategy. The adoption of a discourse of shared responsibilities by the Federal Government within this strategy marked a significant change of direction, explicitly moving away from a regulatory approach and stressing the need for broad participation:

> Climate change is relevant to all parts of society. Only a concerted and co-ordinated effort by all parties will achieve an effective response.

The widening scope of climate change responsibilities included local government, communities and individuals. However, most attention was focused on the *Greenhouse Challenge*, a voluntary programme in which large industries and industry associations undertake cost-effective emissions reductions in return for 'green' publicity. This allocation of responsibilities was seen as crucial for the Federal Government:

> The success of the Greenhouse Challenge depends on a continuing level of contribution from industry that is both real and significant... that means, in turn, that we in government have more flexibility in developing our national greenhouse response.

Despite concerns over the narrow focus of *Greenhouse Challenge*, this approach has been supported by environmental groups, state and local governments. Indeed, it was primarily these organizations who suggested that the *Greenhouse Challenge* programme be extended to small and medium sized businesses and who have initiated similar schemes at state and local levels.

In 1996–97 a more substantial reworking of domestic greenhouse policy was undertaken with the review of the NGRS (NGAP 1996; Intergovernmental Committee on Ecologically Sustainable Development (ICESD) 1997a, b), and the launch of the new National *Greenhouse Strategy* which came into effect in 1998. Throughout this process, the notion of sharing responsibilities across stakeholders and within the community remained central:

> The engagement of stakeholders and the community in identifying and assessing opportunities for appropriate and effective greenhouse response actions is essential to the development of an effective, credible and broadly based Strategy. The new Strategy will reflect a strong commitment to cooperative approaches to address greenhouse issues, utilizing partnerships amongst governments, and between governments and the community.

In order to give state governments a sense of ownership, the development of the new strategy involved allocating key sectors to different states, for example New South Wales were given the task of policy development for the energy sector, while Queensland took on agriculture. The involvement of local government has been promoted through the development of the *Cities for Climate Protection* programme in which participants attempt to reduce the emissions of greenhouse gases from the local area. This programme, run by the International Council for Local Environmental Initiatives, was initially championed by Newcastle City Council and became part of domestic policy in late 1997, when the Federal Government announced its intention to fund its development within Australia. The principle means of fostering the participation of industry within the new strategy remain voluntary, for example the expansion of the *Greenhouse Challenge* programme and continued development of energy efficiency standards. Community participation was effected through two consultation processes, one for suggestions of measures to be included in the draft strategy released in March 1997 and one after its release, as well as by the provision of information in order to persuade people to take action. Neither consultation was a success, with those involved resigned to the fact that such processes tend to attract those that have been paying close attention to the process , and the extent to which the public have taken voluntary action on the basis of information provided is far from clear. Despite evidence of public concern for the issue of climate change, there has been little involvement in the political processes through which risks and responsibilities are delimited. In part this reflects the views of policy makers, who see climate change as an issue which is too scientifically and politically complex to involve lay publics. It also reflects public understanding of climate change risk as something over which they have little control and which should more appropriately be addressed by governments and businesses. This suggests that it will not be in the light of further scientific evidence and disputes over risk that a civic subpolitics of climate change will emerge, as Beck's thesis suggests, but on the basis of providing explicit means through which people feel they can collectively respond and when the responsibilities of other actors and institutions are acknowledged.

Despite the new emphasis on partnership, many of the specific measures contained within the *National Greenhouse Strategy* bear a remarkable similarity to those included in the NGRS and *Greenhouse 21C.* Most, such as the provision of information, regulation, incentives and funding, rely heavily on both federal and state governments for their implementation. While the shift towards partnerships and participation within the climate change policy process has been profound during the 1990s, it has not removed responsibility from the nation-state for shouldering the lion's share of emission reduction responsibilities, and for ensuring that the costs of such action do not fall unduly on particular regions or sectors. The shift in emphasis towards explicit partnership approaches has taken place in the wider context of a weak ecologically modernist reading of climate change possibilities in which the state has continued to privilege the usual businesses. The need to take responsibility for effecting significant changes in the relations between energy use, demand and the economy has been sidelined. Indeed, by appearing willing to undertake some level of responsibility for climate risks, the resource and energy sector has offset more stringent obligations that could have resulted from the implementation of regulations and fiscal measures. The state has played a key role in the politics of climate risks at regional and local levels, and in some cases this has allowed more radical discourses of climate responsibilities to take hold.

For example, state legislation enacted in New South Wales in 1995 requires energy suppliers to reduce emissions of greenhouse gases by 20 per cent by 2005, which has in turn led to investment in renewable energy and the development of 'green energy' schemes for householders. Also in New South Wales, Newcastle City Council has implemented various energy efficiency measures and invested in renewable energy in partnership with private companies. In these cases, the conflicts of accountability wrought by climate change risk have led to new discourses and material practices with respect to energy use as 'debates and conflicts which originate in the dynamic of risk society are... superimposed on interest organizations, the legal system and politics'. While such examples are few and far between, they illustrate that interests do not remain static and can not be considered as given as risk politics take shape. In some places, climate risks have created discourses which attempt to move beyond the spatial and temporal referents of modernity in constructing and delimiting responsibilities for climate change, and in so doing the nature of environmental and social justice.

While the nation-state is responsible for legitimating climate risks and for their alleviation this is a task it can not complete without addressing the source of risks, in this case primarily the use of energy, and without the involvement of the institutions and agents responsible for that use: industry and the community. However, this does not create divisions between the formal political system and subpolitical arenas, rather climate change has provided a new issue in which the interdependencies between these arenas have become re-articulated through discourses of shared responsibilities for risk. These links have been reshaped, as new actors with claims to legitimacy and expertise enter the policy network, and renegotiations of the respective obligations of governments, industries and the community take place. While this process of renegotiation has led to some new discourses of obligations and responsibilities, in the main these remain tied to the temporal and spatial horizons of industrial society.

The issue of climate change has provoked significant conflicts over the accountability of certain places and peoples for the risks imposed on future generations and distant strangers. Issues of blame, responsibility, compensation and obligation have pervaded international negotiations and domestic debates, and have been incorporated into the ensuing agreements and institutions as new discourses and practices are constructed and contested in attempts to legitimate and mitigate climate risk. Beck's supposition that the scale, scope and nature of contemporary environmental risks poses novel challenges for the 'relations of definition' within modernity can be upheld in this case. In drawing attention to the ensuing social and political consequences, Beck's thesis highlights how the cultural and institutional matrix within which risk politics take place constructs and is constructed by contemporary environmental risks. This provides a welcome departure from approaches which view social and political processes as responses to 'given' risks, existing in and of nature in isolation from society. Beck's work also points to the significance of understanding how particular discourses and practices through which risks, interests and obligations are defined gain dominance, and the social structures and interests which such definitions of environmental values maintain and construct. In addition, he signals the importance of considering how sub political institutions are involved in the politics of risk. In Australia, the politics of climate change risk do not take place only within the formal political system, but spill out across a wider policy network as organizations, institutions and individuals struggle to define and contest liability, obligations and responsibility. These insights, drawn from Beck's thesis, have shed considerable light on the ways in which climate change risk and politics take shape within Australia. However, the argument that risk society heralds a new political dynamic, in which a division emerges between an impotent formal political system and various subpolitical arenas, is difficult to sustain in this case. Rather, Beck's alternative vision of the relation between politics and subpolitics, where the formal political system retains power and influence but is thrown into coalitions with subpolitical institutions and actors, seems more appropriate. However, there are several points raised in this analysis which suggest that Beck's thesis does not adequately capture the political dynamics of contemporary risk issues.

The significance of the 'subpolitical' arena for the pursuit of politics is not new. Some 'subpolitical' arenas, notably the resource industries and local communities in resource-dependent regions, have had a long and involved role in Australian politics. Climate change provides a new arena in which these concerns have been articulated, contested and reconstructed. For example, international climate obligations have been seen as warranted by the Federal Government only where unacceptable costs are not imposed unfairly on some economies. This position reflects the interests of the resource and energy industries, as well as some parts of local, state and national governments, who have, since the initiation of the COP, coalesced around the argument that the differentiation of responsibilities internationally and the participation of developing countries is in the national interest. In contrast, environmental groups, scientific organizations, and other parts of local, state and national government have articulated shared discourses that climate change risks create responsibilities to prevent harm to vulnerable societies, the non-human world and future generations. These new concepts of liability and obligation have been resisted by the resource-based coalition and redefined within the horizons of industrial society – the nation-state, the individual, and the market.

In creating obligations and sharing responsibilities in relation to climate risks, the political system is not left devoid of power vis-a'-vis new sub-political arenas and processes. Rather, the process of delimiting risks and responsibilities takes place across the policy network as existing coalitions are redefined and new coalitions take hold. In Australia, the state retains a key role in managing, controlling and legitimating climate risk while also 'strengthening its interaction and dependence on both the private sector and civil society' through new discourses of partnership. Gandy argues that in the light of environmental risks, the capacity of the state for intervention is circumscribed as it becomes 'increasingly dependent on other structures and organizations dispersed through society'. However, these other structures and organizations can include different parts of the formal political system at local and state level. While this is partly a result of Australia's federal structure, the implementation of climate change policy elsewhere will also depend on action taken by lower tiers and different branches of government as well as private companies and individuals. Furthermore, the role of the state is not governed by some determinate and finite notion of capacity, but rather through negotiations in which state and subpolitical actors and institutions mutually define their respective roles. As Hajer argues, environmental politics is an argumentative struggle in which 'actors not only try to make others see problems according to their views but also seek to position other actors in a specific way'. In the politics of climate risk in Australia, the dominant resource-based coalition has sought to restrict the intervention of the state in determining climate responsibilities through regulatory or fiscal measures by promoting voluntary measures. Furthermore, this coalition has argued that responsibilities should be determined on a local or regional scale, taking into account the costs of taking action on climate change on particular regions and industries. Their success is witnessed both in the *National Greenhouse Response Strategy* and *National Greenhouse Strategy*, which emphasize the need to take into account the impacts of action on particular regions and sectors, and in recent legislation which has sought to reign back the influence of the Federal Government on a range of environmental issues, including climate change. The discourse of partnerships, between the state, industry, and community, has arisen not because of a lack of power within the formal political system, but through an argumentative struggle across the policy network in which the resource-based coalition, including state and non-state actors, has sought to channel the influence and control of the state to serve their interests. These negotiations have been conducted within the boundaries of a weakly ecological modernist reading of climate obligations, where responsibilities for climate risk are confined to the economically efficient actions of governments, industries, and concerned individuals. Although some innovative responses to climate change risks have been created, these have been limited in scale and scope, touching only a fraction of the energy economy of Australia, and leaving most questions of need and demand aside. In this light, suggestions that the evolution of subpolitics will effect progressive socio-environmental change must be treated with caution. Any environmental critique of modernity resulting from risk society has made little headway in Australia, where the challenges that climate risks pose have served to reinforce existing relations of definition and relations of production.

References

Arnell N, Cannell MG, Hulme M et al. (2002) The consequences of CO2 stabilization for the impacts of climate change. *Climatic Change*, 53:413–446

Cook SM, Glass RI, LeBaron CW, Ho MS (1990) Global seasonality of rotavirus infections. *Bulletin of the World Health Organization*, 68:171–177

Martens P, Kovats RS, Nijhof S et al. (1999) Climate change and future populations at risk of malaria. *Global Environmental Change—Human and Policy Dimensions*, 9:S89–S107

Rogers DJ, Randolph SE (2000) The global spread of malaria in a future, warmer world. *Science*, 289:1763–1766.

Warrick RA, Barrow EM (1991) Climate change scenarios for the UK. *Transactions of the Institute of British Geographers*, 16:387–399.

7

Global Climate Change and Health Risks

Humans are accustomed to climatic conditions that vary on daily, seasonal and inter-annual time-scales. Accumulating evidence suggests that in addition to this natural climate variability, average climatic conditions measured over extended time periods (conventionally 30 years or longer) are also changing, over and above the natural variation observed on decadal or century time-scales. The causes of this climate change are increasingly well understood. Climatologists have compared climate model simulations of the effects of greenhouse gas (GHG) emissions against observed climate variations in the past, and evaluated possible natural influences such as solar and volcanic activity. They concluded that "... there is new and stronger evidence that most of the warming observed over the last 50 years is likely to be attributable to human activities".

The Third Assessment Report of the IPCC estimates that globally the average land and sea surface temperature has increased by 0.6±0.2°C since the mid-19th century, with much of the change occurring since 1976. Warming has been observed in all continents, with the greatest temperature changes occurring at middle and high latitudes in the Northern Hemisphere. Patterns of precipitation have also changed: arid and semi-arid regions are apparently becoming drier, while other areas, especially mid-to-high latitudes, are becoming wetter. Where precipitation has increased, there has also been a disproportionate increase in the frequency of the heaviest precipitation events. The small amount of climatic change that has occurred so far has already had demonstrable effects on a wide variety of natural ecosystems.

Climate model simulations have been used to estimate the effects of past, present and likely future GHG emissions on climate changes. These models are primarily based on data on the heat-retaining properties of gases released into the atmosphere from natural and anthropogenic (man-made) sources, as well as the measured climatic effects of other natural phenomena, as described above. The models used by the IPCC have been validated by "back-casting"—that is, testing their ability to explain climate variations that already occurred in the past. In general, the models are able to give good approximations of past patterns only when anthropogenic emissions of non-GHG air pollutants (particulates, dust, oxides of sulfur, etc.) are included along with natural phenomena. This emphasizes that (i) the models represent a good approximation of the climate system; (ii) natural variations are important contributors to climatic variations, but cannot adequately explain past trends on their own; and (iii) anthropogenic GHG emissions are an important contributor to climate patterns, and are likely to remain so in the future.

Considering a range of alternative economic development scenarios and model parameterizations, the IPCC concluded that if no specific actions were taken to reduce GHG emissions, global temperatures would rise between 1.4 and 5.8°C from 1990 to 2100. The projections for precipitation and wind speed are less consistent in terms of magnitude and geographical distribution, but also suggest significant changes in both mean conditions and in the frequency and intensity of extreme events.

Estimating the Effects of Climate Change on Health

Human health is sensitive to temporal and geographical variations in weather (short-term fluctuations in meteorological conditions) and climate (longer-term averages of weather conditions). Weather has not historically been considered as subject to alteration by human actions, although its effects may be lessened by adaptation measures. While adaptation is also a very important determinant of the health consequences of climate change, the effect of anthropogenic GHG emissions on climate means that climate change can in principle be considered a risk factor that could potentially be altered by human intervention, with associated effects on the burden of disease.

The effects of GHG emissions on human health differ somewhat from the effects of other risk factors in that they are mediated by a diversity of causal pathways and eventual outcomes, typically long delays between cause and effect, and great difficulties in eliminating or substantially reducing the risk factors. An additional challenge is that climate change occurs against a background of substantial natural climate variability, and its health effects are confounded by simultaneous changes in many other influences on population health. Empirical observation of the health consequences of long-term climate change, followed by formulation, testing and then modification of hypotheses would therefore require long time-series (probably several decades) of careful monitoring. While this process may accord with the canons of empirical science, it would not provide the timely information needed to inform current policy decisions on GHG emission abatement, so as to offset possible health consequences in the future. Nor would it allow early implementation of policies for adaptation to climate changes, which are inevitable owing to both natural variations and past GHG emissions. Therefore, the best estimation of the future health effects of climate change will necessarily come from modelling based on current understanding of the effects of climate (not weather) variation on health from observations made in the present and recent past, acknowledging the influence of a large range of mediating factors.

Since the early 1990s, IPCC Working Group II has collated some of the accumulating predictions of climate effects on health. The health effects of climate variability and change have also been reviewed by a national committee in the United States of America and, more recently, in a book by the World Health Organization (WHO). As yet, however, there has been no concerted attempt to integrate these various research findings into a single standardized estimate of the likely net health effects of climate change, nor to estimate the possible health gains associated with different mitigation and amelioration strategies. In addition to the uncertainties of future climate projections, there are several obstacles to achieving this aim.

- Not all of the probable health outcomes have been modelled, often due to lack of parameterization data and the complexity of causal pathways. Modelling efforts so far have tended to concentrate on those causal relationships that can more easily be modelled, rather than those with the potentially greatest effects (e.g. extreme temperatures on cardiovascular disease mortality, rather than sea-level rise on the health of displaced populations).

- Little emphasis has been given to the validation of models relating climate change to health. Validation would provide a basis for making uncertainty estimates around projections, and would afford an objective criterion for choosing between different models or modelling approaches.
- Adaptations to climate change (i.e. autonomous or planned responses that reduce the vulnerability of populations to the consequences of climate change) are often not addressed.
- Interactions between the effects of climate change and other changes to human populations (e.g. investment in health infrastructure, level and equity of distribution of wealth) are seldom explicitly estimated.
- The various disease-specific models invariably generate outputs in different units, which may be only indirectly related to disease burden (e.g. populations at risk of disease transmission, rather than disease incidence). This hampers estimation of the aggregated health impacts of different scenarios.
- Little effort has previously been directed to describing and understanding the geographical variations in likely impacts.

The first four obstacles are likely to be at least partially addressed in the future, as disease-specific models become more sophisticated and, perhaps more importantly, through the accumulation of greater quantities of reliable data for model parameterization and testing.

Risk Factor Definition and Measurement

Definitions of Risk Factor and Exposure Scenarios

The risk factor was defined as current and future changes in global climate attributable to increasing atmospheric concentrations of greenhouse gases (GHGs).

Composite climate scenarios are adopted instead of the (more preferable) continuous measurements of individual climate variables because (i) climate is a multivariate phenomenon, including temperature, precipitation, wind speed, etc., and therefore cannot be measured on a single scale; (ii) climate changes will vary significantly with geography and time: these are not fully captured in global averages of climate variables; and (iii) all aspects of climate are likely to be altered by GHG levels in the atmosphere.

The exposure categories considered here are global climate scenarios resulting at specified points in time over the coming half-century from:

1. unmitigated emissions trends, that is, approximately following the IPCC IS92a scenario;
2. emissions reduction, resulting in stabilization at 750ppm CO_2 equivalent by the year 2210 (s750);
3. more severe emissions reduction, resulting in stabilization at 550 ppm CO_2 equivalent by the year 2170 (s550);
4. average climate conditions for 1961–1990, the World Meteorological Organization (WMO) climate normal (baseline).

Although future GHG emissions are inherently uncertain, the unmitigated emissions scenario adopted here was, until recently, the IPCC mid-range projection, and was very widely used in climate impact modelling. The stabilization categories used here represent plausible, though economically and technically challenging, projections that are dependent on there being major efforts to curtail emissions.

We do not attempt to estimate all health outcomes of specific policy/ development pathways through which these, or other, GHG levels could be achieved: for example, compliance with the Kyoto Protocol of the United Nations Framework Convention on Climate Change (UNFCCC), or of the world following one or other of the IPCC Special Report on Emissions Scenarios (SRES)—both of which also include descriptions of alternative future global socioeconomic development scenarios. The costs or additional benefits of specific interventions to achieve this reduction are artificially separated from the resulting health benefits. For such a distal risk factor, this separation of intervention from exposure and resulting health consequences may potentially introduce inconsistencies: for example, the economic changes necessary to achieve GHG stabilization are more consistent with some projections of levels and distribution of population and GDP than others. These socioeco-nomic factors are themselves likely to effect disease rates, potentially in interaction with climate. Integrated assessment of all effects of interventions would be conceptually more consistent, but this has not been attempted here, since it would introduce an additional layer of uncertainty and assumptions into the models and has previously only been explored for a few health outcomes.

The choice of baseline or "theoretical minimum" exposure follows the WMO and IPCC practice of using the observed global climate normal for 1961–1990 as a reference point. Alternative baselines, such as pre-industrial climate, are not used, because of the absence of a published consensus among climatologists on definitions of an appropriate time period and the relative roles of anthropogenic and natural influences before 1961–1990. This choice of a 1961–1990 baseline will therefore generate relatively conservative estimates of change in exposure and associated disease outcomes, as it does not address any human-induced climate change that occurred before this period.

The approach here treats climate change as a slowly evolving and continuous exposure, with the majority of disease models linked to those changes for which climate models make the most consistent predictions: gradual changes in temperature and, to a lesser extent, precipitation. This is again a limited approach. It is very likely that climate change will also increase the frequency of extreme conditions, with likely effects on health. However, quantitative estimates of increased frequency have only very recently become available for some measures of extremes, and are not yet available for different GHG scenarios. The consequences of such changes are modelled here only in the context of inland flooding, but they could potentially be applied to other health end-points in the future.

Finally, there is some concern that disruption of the climate system may pass critical thresholds, resulting in abrupt rather than gradual changes and associated rapid impacts on health. There is no consensus on the probability of such events, and they have therefore not been included in any published health outcome assessment studies. However, they should be borne in mind as a "worst-case" scenario.

Methods for Estimating Risk Factor Levels

Projections of the extent and geographical distribution of climate change were generated by applying the various emissions scenarios described above to the HadCM2 global climate model (GCM) of the Hadley Centre in the United Kingdom of Great Britain and Northern Ireland. This is one of several alternative GCMs used by the IPCC; it generates projections of changes in temperature and other climate properties which have been verified by back-casting , and which lie approximately in the middle of the range generated by alternative models.

The HadCM2 model generates estimates of the principal characteristics of climate, including temperature, precipitation and absolute humidity, for each cell of a global grid at resolution 3.75° longitude by 2.5 ° latitude. As for most climate change models, HadCM2 generates daily projections representing both long-term trends and the degree of natural climate variability, but not necessarily its specific temporal pattern (i.e. the models do not accurately predict the climate of specific days or months). In order to account for such natural variability, the outputs that are most commonly used for modelling consequences of climate change are monthly means for average 30-year periods centred on the 2020s, 2050s, 2080s, etc. The baseline climate describes the same properties for the land surface of the world at 0.5 °*x* 0.5° resolution.

The climate model projections describe forecast changes in global climate conditions. Therefore, we did not attempt to estimate the "exposure prevalence": the entire world population was assumed to be exposed to one or other global climate scenario (i.e. exposure prevalence =100%). However, it should be noted that the climate scenarios incorporate geographical variation in both current climate (e.g. the lower temperatures in higher latitudes) and projected climate change (e.g. more rapid and intense warming is predicted in high northern latitudes than elsewhere). Different populations will therefore experience different climate conditions under any one climate-change scenario.

Uncertainties in Risk Factor Levels

Two major sources of uncertainty surround the forecasting of future climate scenarios: (i) uncertainty in changes in factors such as population, economic growth, energy policies and practices on GHG emissions; and (ii) uncertainties over the accuracy of any climate model in predicting the effects of specified emissions scenarios on future climate in specific locations, against the background of substantial natural climate variability over time, and in space (i.e. downscaling). Climatologists have only very recently provided probabilistic measurements of uncertainty incorporating one or both of these sources, and there is still debate over the reliability and utility of such measures. They have not previously been applied in impact studies , and were therefore not used in this assessment. More importantly, however, there remains considerable uncertainty over the accuracy with which any single model can predict future climate. This is usually addressed by using outputs from a range of models, from independent groups. This was not possible here, as the particular GHG stabilization scenarios have only been applied, and fed through to estimates of likely consequences, for the HadCM2 model.

For this analysis, we did not address the first and last sources of uncertainty explicitly, and assumed that it is incorporated in the various alternative exposure scenarios, reflecting different trajectories of GHG emissions. The second source of uncertainty was partially addressed by using 30-year averages of climate conditions, which helps to "smooth out" the effects of natural climate variability. Further, the single model used was run with slight variations in initial conditions, allowing the calculation of an "ensemble mean". Four runs were used to generate an ensemble mean for the unmitigated emissions scenario. However, only single climate runs were available for the stabilization scenarios. Therefore, the climate scenarios associated with those emissions scenarios are more uncertain.

Here, the stabilization scenarios were run on a suite of 14 simple climate models, making different plausible assumptions about climate sensitivity to GHG emissions using the simplified COSMIC climate model described by Schlesinger and Williams. This illustrates the degree of variation between the projections for future temperature and precipitation patterns. (Note that projections for precipitation vary more between models than does temperature— therefore models that rely on precipitation estimates from single scenarios have an additional component of uncertainty.)

Risk Factor–Disease Relationship

Outcomes Included

The health outcomes addressed here were selected on the basis of observed sensitivity to temporal and geographical climate variation, importance in terms of mortality and/or burden of disease and availability of quantitative global models (or feasibility of constructing informative models in the time available). More detail on evidence for causality and quantitative estimation methods for each health outcome is given below.

Climate change is by and large a relatively distal risk factor for ill-health, often acting through complex causal pathways which result in heterogeneous effects across populations. There is, therefore, a series of additional likely outcomes that have not yet been formally modelled. They include the potential health consequences of climate change on:

- changes in pollution and aeroallergen levels;
- the rate of recovery of the ozone hole, affecting exposure to ultraviolet radiation ;
- changes in the distribution and transmission of other infectious diseases, particularly other vector-borne diseases, geohelminths and rodent-borne diseases, and possible emergence of new pathogens;
- the distribution and abundance of plant and livestock pests and diseases, affecting agricultural production;
- the probability of crop failure through prolonged dry weather and famine, depending on location and crisis management;
- population displacement due to natural disasters, crop failure, water shortages; and
- destruction of health infrastructure in natural disasters.

Methods for Estimating Risk Factor–Disease Relationships

Various methods have been developed for the quantitative estimation of health outcomes of climatic change. It is not yet feasible to base future projections on observed long-term climate trends, for three reasons: (i) the lack of standardized long-term monitoring of climate-sensitive diseases in many regions; (ii) methodological difficulties in measuring and controlling for non-climatic influences on long-term health trends; and (iii) the small (but significant) climate changes that have occurred so far are an inadequate proxy for the larger changes that are forecast for coming decades.

Instead, estimates are based on observations of the effects of shorter-term climate variation in the recent past (e.g. the effects of daily or inter-annual climate variability on specific health outcomes) or the present (e.g. climate as a determinant of current disease distribution), or on specific processes that may influence health states (e.g. parasite and vector population dynamics in the laboratory, determining the transmission of infectious diseases). These quantitative relationships were then applied to future climate scenarios. Such an approach makes the important assumption that such associations will be maintained in the future, despite changes in mediating factors such as socioeconomic variables, infrastructure and technology. This introduces significant uncertainty, and possibly bias, in the estimates.

The extent and type of modelling applied to different health effects vary considerably. Consequently, several outcomes can only be estimated by crude adaptation of the outputs of available models. For example, some of the predictive models generate health-relevant outputs that do not correspond directly to categories of disease states used in the Global Burden of Disease (GBD) study. These include the incidence of deaths and injuries due to, specifically, floods (rather than injuries due to all causes), or populations at risk of hunger or malaria infection (rather than prevalence of malnutrition, or incidence of clinical malaria). Currently, there are spatial resolution differences between models, which is not ideal. These relate to how the models account for geographical variation in: (i) baseline climate (potentially differentiated to 0.5 ° globally, or higher resolutions for some regions); (ii) climate change (usually at the level of the GCM projections: 3.75° longitude by 2.5° latitude); and (iii) aggregation of the final results (occasionally according to regions other than subregions, depending on the purpose of the original model).

Assumptions

Simplifying assumptions have been made to facilitate clear definition of scenarios and associated consequences.

Different Mechanisms for Reducing GHG Emissions

The alternative GHG emissions scenarios outlined above could be achieved through an almost infinite variety of changes to economic and social development and energy use policies. They include relatively direct mechanisms which may be *negative*, such as the potential negative effects of GHG emissions policies on economic development, personal wealth and vulnerability to disease , or *positive*, such as reduction in the levels of ozone and other outdoor and indoor air pollutants. They may also act through more complex routes, for example, avoiding production of aeroallergens in CO_2-enriched environments.

Population Growth

The models described below estimated the relative per capita incidence of specific health outcomes under the different climate scenarios. The size and distribution of current and future populations therefore affect the relative risk estimates either (i) where the climate hazard is not evenly distributed geographically throughout the region, or (ii) where population is an integral part of the model—for example, in the risk-of-hunger model, where population size has an influence on food availability per capita. In adjusting the relative risks for these population effects, the distribution of future populations was estimated by applying the World Bank mid-range estimate of population growth either at the national level (for malnutrition), or to a 1°¥1° resolution grid map of population distribution for all other outcomes.

Modifying Factors: Adaptation and Vulnerability

Factors such as physiological adaptation, technological and institutional innovation and individual and community wealth will influence not only the exposure of individuals and populations to climate hazards, but also the associated hazards. For some assessments, simpler modifying factors are integrated into the models for both present and future effects. For example, estimates of changes in the number of people at risk of hunger incorporate continental projections of economic growth, affecting capacity to buy food. Other models incorporate the effects of existing modifiers when defining current climate–disease relationships, such as estimates of the global distribution of malaria based on current climate associations. Such models implicitly capture the *current* modifying effects of socioeconomic and other influences on climate effects, but do not attempt to model *future* changes in these modifiers. Finally, some models make no estimate of such modifying influences in either the present or future; for example, models that estimate future changes in the geographical range which is climatically suitable for malaria transmission, and associated populations at risk. To generate consistent estimates in this analysis:

- we attempted to account for current geographical variation in vulnerability to climate, where not already incorporated into the predictive models.
- we attempted to account for future changes in disease rates due to other factors (e.g. decreasing rates of infectious disease due to technological advances/improving socioeconomic status), and for changes in population size and age structure (e.g. potentially greater proportion of older people at higher risk of mortality related to cardiovascular disease in response to thermal extremes). This was addressed by calculating only relative risks under alternative climate change scenarios, which should be applied to GBD projections of disease rates and population size and age structure. The GBD projections take into account the effects of changing GDP, "human capital" (as measured by average years of female education), and time (to account for trends

such as technological development) on the overall "envelope" of cause-specific mortality and morbidity for diseases affected by climate change.

- all quantitative estimates of the health effects of climate change were based on observed effects of climate variations either over short time periods, or between locations. They therefore made the important assumption that these relationships are also relevant to long-term climate change. To avoid unrealistic extrapolation of short-term relationships, we included consideration of mechanisms by which climate-health relationships may alter over time (i.e. adaptation). We considered whether each disease in turn was likely to be significantly affected by biological adaptation (i.e. either behavioural, immunological or physiological) and/or by generally improving socioeconomic conditions (i.e. increasing GDP). In each case we defined appropriate adjustments to the relative risk estimates, in line with published studies. The different factors for each health outcome were then applied to the same projections of future GDP and changing climate, to adjust the relative risks over the time course of the assessment. There is, however, substantial uncertainty over the most likely degree of adaptation under different conditions. This was reflected in the uncertainty estimates for the relative risks for each disease. Quoted uncertainty estimates therefore describe uncertainties around climate change predictions, about current exposure–response relationships, and around the degree to which these are likely to be maintained in the future. Since, to date, these have not been formally modelled, they were generated here by qualitative assessment in collaboration with the original modelling group. The uncertainty estimates should therefore be interpreted with caution.
- we ignored more complex aspects of future vulnerability. Whereas projected trends for average income (which is included in estimating baseline rates and, where possible, relative risks) are broadly positive, other factors may have opposite effects. These include income distribution, maintenance of disease surveillance, control and eradication programmes, technological change and secondary or threshold effects, such as the protective effect of forests in reducing the frequency and intensity of flooding.

- we made no attempt to estimate the effects of actions taken specifically to adapt to the effects of climate change (e.g. the upgrading of flood defences specifically to cope with sea-level rise attributable to climate change). Therefore, our estimates represent a "business-as-usual" scenario of the health effects associated with global climate change.

Estimation for Different Time Points

As stated above, climate model outputs are usually presented as averages over 30-year periods, for example, centred on 2025 and 2055. In order to generate estimates for any specific year, we defined 1990 (i.e. the last year of the baseline climate period) as year 0. Central, lower and upper estimates of relative risks were calculated for the 2020s and the 2050s, using models described elsewhere. Quoted estimates for the years 2001, 2005, 2010 and 2020 were estimated by linear regression against time between 1990 and 2025. Estimates for 2030 were estimated by linear regression between 2025 and 2055. Using this method, our choice of 1990 as the baseline year rather than the middle of the 1961–1990 period led to conservative estimates of health consequences, particularly in the near future.

Risk Reversibility

In the context of this assessment, risk reversibility describes the proportion of the health consequences that would be avoided if the population were shifted to a different exposure scenario. Given the definition of exposure scenarios, complete avoidance of climate change (i.e. populations exposed to baseline climate conditions rather than unmitigated climate change) would avoid all of the health consequences that we have described. Risk reversibility would therefore be 100% in this case.

Criteria for Identifying Relevant Studies and for Estimating Strength of Evidence

In identifying relevant studies, we have considered all publications reviewed in the IPCC Third Assessment Report , as well as those found in more recent literature searches. However, as the field is relatively new and expanding rapidly, we have also used some studies that are either in-press or submitted for publication (material available on request).

There are still relatively few models that link climate change models to quantitative global estimates of health or health-relevant outcomes (e.g. numbers of people flooded or at risk of hunger). Where global models did not exist, we have extrapolated from models that make local or regional projections. Methods for extrapolating to the subregions are described separately for each disease. Where there was more than one published model for a particular outcome, decisions on model selection were based as much as possible on validation against historical/ geographical patterns. We excluded models: (i) that have been shown to be significantly less accurate than equivalent models in predicting historical or geographical trends; or (ii) that have been superseded by later models by the same group; or (iii) that are based on unrealistic biological assumptions; or (iv) that cannot be plausibly extrapolated to a wider area. Given the very limited means for model validation, these results involved choices made in this work among models and resulted in very large uncertainty.

As climate change impact assessment is, at this stage, predominantly a model-based exercise, the assessment of strength of evidence for causality is necessarily indirect. It is based on two considerations: (i) the strength of evidence for the current role of climate variability in affecting the health outcome, and (ii) the likelihood that this relationship between climate and health will be maintained through the process of long-term climate change.

Several of the uncertainties involved in this exercise, particularly those relating to climate modelling and those around the quantitative relationships between climate and health, should decrease through improvements in modelling, and more importantly, through better empirical research. However, estimation of climate change effects is currently a predictive exercise, based on some form of indirect modelling (e.g. analysis of temporal or geographical variation in climate), rather than direct experience (i.e. of a change process that has previously occurred). It is therefore possible that some of the health outcomes listed above may not respond in the predicted manner. On the other hand, we do not yet have direct experience of the full range of health outcomes that may be associated with exposure.

The analyses described here were specifically to estimate the future disease burden, which may be avoided by climate change mitigation policies. However, it must be emphasized that these burdens may also be reduced by adaptation interventions to reduce the vulnerability of populations. Given the "exposure commitment" (unavoidable climate change due to past GHG emissions), and the large gap between even plausible mitigation scenarios and the baseline climate, adaptation strategies are essential to the goal of reducing adverse health effects. These include early warning systems and defences for protection from natural disasters, and improved health infrastructure (e.g. water and sanitation, control programmes for vector-borne diseases) to reduce the baseline incidence of climate-sensitive diseases.

Direct Effects of Heat and Cold on Mortality

An association between weather and daily mortality has been shown consistently in many studies in diverse populations. The effect of extreme temperature events (heat-waves and cold spells) on mortality has also been well described in developed countries. However, temperature-attributable mortality has also been found at "moderate" temperatures.

Causality is supported by physiological studies of the effects of very high or very low temperatures in healthy volunteers. High temperatures cause some well described clinical syndromes such as heat-stroke. Very few deaths are reported as directly attributed to "heat" (International Classification of Diseases, ninth revision in most countries. Exposure to *high* temperatures increases blood viscosity and it is therefore plausible that heat stress may trigger a vascular event such as heart attack or stroke. Studies have also shown that elderly people have impaired temperature regulation. Clinical and laboratory studies indicate that exposure to *low* temperatures causes changes in haemostasis, blood viscosity, lipids, vasoconstriction and the sympathetic nervous system. The strongest physiological evidence is therefore for cardiovascular disease.

Population-based studies also provide evidence that environmental temperature affects mortality due to both cardiovascular and respiratory disease. The best epidemiological evidence is provided by time-series studies of daily mortality. These methods are considered sufficiently rigorous to assess short-term associations (days, weeks) between environmental exposures and mortality, if adjustment is made for longer-term patterns in the data series, particularly the seasonal cycle and any long-term trends, such as gradually decreasing mortality rates over decades.

The effect of a "hot" day is apparent for only a few days in the mortality series; in contrast, a "cold" day has an effect that lasts up to two weeks. Further, in many temperate countries, mortality rates in winter are 10–25% higher than death rates in summer. The causes of this winter excess are not well understood. It is therefore plausible that different mechanisms are involved and that cold-related mortality in temperate countries is related in some part to the occurrence of seasonal respiratory infections.

Although the physiological evidence for causality of an effect of temperature on mortality is greatest for cardiovascular, followed by respiratory, diseases, temperature has been shown to affect all-cause mortality in areas where cardiovascular disease rates are relatively low and infectious disease mortality is relatively high. Many studies report the seasonal patterns of infectious disease in developing countries, but the role of temperature is not well described. It is likely that seasonal rains influence the seasonal transmission of many infectious diseases. High temperatures encourage the growth of pathogens and are associated with an increased risk of diarrhoeal disease in poorer populations.

Studies have also described mortality and morbidity during extreme temperature events (heat-waves). However, these events are, by definition, rare. It is therefore difficult to compare heat-waves in different populations and for different intensities. The studies that have been used to describe the effects of heat-waves also use different methods for estimating the "expected" mortality, which makes comparison difficult. An assessment of the health consequences of climate change on thermal stress requires estimation of past and future probabilities of extreme temperature events. The current methods of assessment use 30-year averages of monthly data, and few scenarios consider change in the frequency or magnitude of extreme events. This is because: (i) suitable methods have not been developed, and (ii) climate model output at the appropriate spatial and temporal resolution is not readily available.

There is little published evidence of an association between weather conditions and measures of morbidity such as hospital admissions or primary care consultations. A study of general practitioner consultations among the elderly in Greater London found that temperature affected the rate of consultation for respiratory diseases but not that for cardiovascular diseases. However, it is not clear how these end-points relate to quantitative measures of health burden.

Estimating the Temperature–Mortality Relationship

A review of the literature was undertaken to identify studies that report relationships between daily temperature and mortality.

The following criteria were used to select studies for deriving the modelled estimates.

- A study that uses daily time-series methods to analyse the relationship between daily mean temperature and mortality.
- A study that reports a coefficient from linear regression which estimates the percentage changes in mortality per degree centigrade changes in temperature above and below a reported threshold temperature.

The best characterized temperature–mortality relationships are those for total mortality in temperate countries. Fewer studies have also looked at the particular causes of death for which physiological evidence is strongest: cardiovascular disease, and to a lesser extent respiratory disease. In this study, we used the specific relationships for cardiovascular disease where these were available (temperate and cold-climate zones), and the general relationships for all-cause mortality for climatic regions where such disease-specific relationships could not be found in the literature review (all tropical populations).

The global population distribution was divided into five climate zones , according to definitions developed for urban areas by the Australian Bureau of Meteorology. The population in the polar zone is small (0.2% of world population) and was excluded.

It was necessary to estimate daily temperature distributions in order to calculate the number of attributable deaths. Daily temperature distributions clearly vary a great deal, even between localities within the same country. However, it was not feasible within this assessment to obtain sufficient meteorological data to estimate daily temperature distributions throughout the world at a fine spatial resolution. Therefore, a single distribution was chosen to represent each climate zone. New daily temperature distributions were then estimated for each climate scenario, by shifting the currently observed temperature distributions by the projected change in mean temperatures for each month, and of the variability of daily temperatures as well as changes in the mean.

Estimating Temperature-attributable Mortality

An exposure–response relationship and threshold temperature (T_{cutoff}) was applied within each climate zone. The average temperature difference above (hot days) and below (cold days) this temperature was calculated for baseline climate and each of the climate scenarios.

The short-term relationships between daily temperature and mortality were used to estimate the annual attributable fraction of deaths due to hot days and cold days for each of the climate scenarios (i.e. annual temperature distributions based on averages over 30 years). Deaths attributable to climate change were calculated as the change in proportion of temperature-attributable deaths (i.e. heat-attributable deaths plus cold-attributable deaths) for each climate scenario compared to the baseline climate. The 1°¥1° resolution grid map of population distribution was then overlaid on the maps of climate zones in a geographical information system (GIS), to estimate the proportion of the population in each subregion who live in each climate zone. The proportional changes in temperature-attributable deaths were therefore calculated by taking the average of the changes in each climate zone represented in the subregion, weighted by the proportion of the subregion's population living within that climate zone.

Adaptation

Acclimatization includes autonomous adaptation in the individual (physiological adaptation, changes in behaviour) and autonomous and planned population-level adaptations (public health interventions and changes in built environment). Acclimatization to warmer climate regimes is likely to occur in individuals and populations, given the rate of change in *mean* climate conditions currently projected by climate models. However, it is uncertain whether populations are able to adapt to non-linear increases in the frequency or intensity of daily temperature extremes (heat-waves). Even small increases in average temperature can result in large shifts in the frequency of extremes.

Few studies have attempted to incorporate acclimatization into future projections of temperature-related mortality , but all studies report that acclimatization would reduce potential increases in heat-related mortality. Our estimates incorporated an assumption regarding acclimatization of the populations to the changing climate that describes this reduced effect. We assumed that the threshold temperature (T_{cutoff}) is increased as populations adapt to a new climate regime, reflecting physiological and behavioural acclimatization that can take place over the time-scale of decades. Changes in T_{cutoff} are region and scenario specific, as they reflect the rate of warming experienced. Therefore, they were assumed to be proportional to the projected change in average summer temperature (Dt_{summer}, equal to the mean of the three hottest months) from the climate scenario. The temperature–mortality relationships were assumed not to change over time; that is, populations biologically adapt to their new average temperatures, but remain equally vulnerable to departures from these conditions. We made no explicit adjustment for an effect of socioeconomic development and technological change on temperature-related mortality.

Short-term Mortality Displacement

Evidence suggests that the increase in mortality caused by high temperatures is partially offset by decreased deaths in a subsequent "rebound" period. This indicates that some of the observed increase in heat-related mortality may be displacement of deaths among those with pre-existing illness, which would have occurred soon in any case. However, this effect has not been quantified for temperature exposures and was not included in the model. The estimates are therefore used to calculate only attributable deaths, but not DALYs, as the estimate of attributable years of life lost was highly uncertain.

In subregions with predominantly temperate and cold climates, reductions in cold-related mortality are likely to be greater than increases in heat-related mortality. Therefore, all climate scenarios show a net benefit on mortality in these subregions, consistent with the IPCC conclusions described above. The effect of the adaptation assumption is to reduce relative risks and therefore net mortality.

Uncertainty Estimates

The principal uncertainty in these estimates, and for all other health effects of climate change, relates to the extrapolation of a short-term climate-health relationship to the long-term effects of climate change. The degree to which this is a reasonable extrapolation relates to the degree to which populations will adapt to changing temperatures, both in terms of reducing the additional mortality attributable to heat, and the possible benefits of avoiding cold deaths. This outcome is unusual, in that the predicted health effects of climate change are negative in some regions, but positive in others. We therefore use slightly different terminology to describe the range of uncertainty around the estimates. The mid-range estimate was given by applying the model described above (i.e. making an adjustment for biological adaptation). The "high-impact" estimate assumes that there was no physiological or behavioural adaptation, and therefore no change in the dose–response relationship over time. This maximizes both positive and negative effects. The low-impact estimate assumes complete adaptation to a changing climate, and therefore no change in the relative risk as the climate changes.

There is therefore a need for further time-series studies, applying a standard approach to populations living in as wide a range of climates as possible. Previous analyses have focused mainly on temperate zones, where the winter effects are greatest, potentially causing over-estimation of the reduced burden attributable to climate change. More analyses of temperature–mortality relationships are therefore required in tropical developing countries.

Such studies should attempt to estimate formally the degree to which adaptation may decrease mortality, and to which observed associations between climate and mortality reflect displaced rather than additional deaths. Finally, there is a need for greater investigation of the health burden of morbidity associated with temperature extremes, including, for example, inability to work in extreme temperatures.

Diarrhoeal Diseases

Diarrhoeal diseases are highly sensitive to climate, showing seasonal variations in numerous sites. This observation is supported by regression analyses of the effects of seasonal and longer-term variation in a limited number of sites. The climate-sensitivity of diarrhoeal disease is consistent with observations of the direct effects of climate variables on the causative agents. Temperature and relative humidity have a direct influence on the rate of replication of bacterial and protozoan pathogens, and on the survival of enteroviruses in the environment. Rainfall may affect the frequency and level of contamination of drinking water.

Quantitative relationships can be defined between climate variations and incidence, which can in turn be directly linked to the outputs of global climate change models. There are, however, challenges and uncertainties in estimating the magnitude of effects.

- The sites from which these relationships were defined cover only a small part of the spectrum of global climate variation. Different relationships may apply at higher or lower temperatures.
- The relative importance of different pathogens and modes of transmission (e.g. via water, food, insects or human–human contact) varies between locations, and is heavily influenced by level of sanitation. As the pathogens are known to vary in their response to climate, this will cause uncertainty in extrapolating temperature

relationships from local studies to other regions with different levels of development.

- Pathogens vary in the severity of clinical symptoms, and the likelihood that they will be reported to health services. Therefore, climate–disease relationships derived only from passive reporting may differ from those based on other methods of surveillance.
- While several studies describe climate effects on particular diarrhoea pathogens, these cannot be used directly to estimate effects on diarrhoeal disease without information on: (i) their relative contribution to overall disease incidence, and (ii) equivalent data on climate sensitivity and relative prevalence for all other diarrhoea pathogens.
- Despite convincing evidence on the effect of extreme rainfall on water-borne outbreaks of diarrhoea, even in highly developed countries , this cannot easily be generalized to the total burden of diarrhoeal disease without information on the relative contribution of such outbreaks to overall diarrhoea incidence.
- Rainfall effects on overall diarrhoea (where observed) are non-linear, and cannot easily be extrapolated to other regions.

In order to minimize these uncertainties, we restricted our estimates to the effect of increasing temperatures on the incidence of all-cause diarrhoea reported to health services (i.e. without attempting to make separate estimates for different pathogens, transmission routes, severities, or for the more complex associations with rainfall). There are important residual uncertainties related to the extrapolation of temperature relationships from specific study sites to others with different temperature regimes and levels of development, and climate effects on reported diarrhoea compared to the true burden of disease. Projections of increasing frequency of extreme wet seasons are very large (e.g. two to five times increase in the regions analysed; Palmer and Ralsanen 2002), and extreme precipitation is associated with increased diarrhoea in both developed and developing countries.

Graded maps of 3.75 ° longitude by 2.5° latitude resolution showing the change in temperature under the alternative scenarios were overlaid with 0.5 °¥0.5° resolution maps of predicted population distributions for the 2020s and 2050s in a GIS. The GIS was used to calculate the average change in exposure (temperature) for each population grid-cell.

Although seasonality of diarrhoeal disease is well recognized, the quantitative relationship between climate and overall diarrhoea incidence has only been explicitly measured in two studies. Both studies described relationships with all-cause diarrhoea, that is, specific pathogens were not differentiated. Checkley et al. used time-series analysis to correlate measurements of temperature and relative humidity against daily hospital admissions at a single paediatric diarrhoeal disease clinic in Lima, Peru for just under 6 years. A total of 57 331 admissions were recorded during the study. The analysis showed a 4% (95% CI 2–5%) increase in admissions for each degree centigrade increase in temperature during the hotter months, and a 12% (95% CI 10–14%) per degree centigrade increase in the cooler months, averaging at an 8% (95% CI 7–9%) per degree centigrade increase over the course of the study. During the 1997–1998 El Niño period there was an additional increase in admissions above that expected on the basis of pre-El Niño temperature relationships, but no association with relative humidity independent of temperature. No rainfall during that period.

In the Checkley et al. study, exposure (climate) data were recorded at local meteorological stations, and can be considered to have negligible measurement error at the population level. The analysis independently controlled for seasonal variations and long-term trends, imparting high confidence to the observed effect of temperature on the outcome recorded. The positive correlation with temperature is also biologically plausible, as a high proportion of diarrhoea cases in many tropical developing countries are caused by bacteria, entamoeba and protozoa , which are favoured by high temperatures. The principle limitations of the study by Checkley et al. are that the outcome recorded may not be representative of climate effects on: (i) less severe disease (i.e. not requiring hospitalization) or more severe disease (diar-rhoeal deaths), or (ii) disease in adults rather than children.

Singh et al. used similar time-series analyses to correlate monthly reported incidence of diarrhoea throughout Fiji against variations in temperature and rainfall, after allowing for the effects of seasonal variation and long-term trend. The study covered the period between 1978 and 1998, with an average of some 1000 reported cases for each of the 228 study months. Reported incidence increased by approximately 3% (95% CI 1.2–5.0%) for each degree centigrade increase in temperature, by 2% (95% CI 1.5–2.3%) per unit increase in rainfall above average rainfall conditions (5¥10^{-5}kg/m^2 per min), and by 8% per unit decrease below average conditions. The pattern is supported by a positive geographical correlation between temperature and incidence in 18 Pacific Island countries. Climate measurements were from a 2.5 °¥2.5 ° cell of a global gridded data set corresponding to Fiji. The use of monthly averages of climate data from this large geographical area may not have reflected the full range of climate exposures of the population; this would have introduced random error and decreased sensitivity. Low rainfall may force use of contaminated water, while high rainfall may contaminate water through flooding. The major limitations of this study are lack of a clear clinical or laboratory definition for diarrhoea, and lack of information on the age distribution of cases.

We are not aware of any similar studies of climate effects on all-cause diarrhoea in developed regions, although studies have been carried out on some subsets of total incidence. Bentham showed that the incidence of food poisoning, usually caused by bacteria, increased by approximately 9% per degree centigrade in England and Wales. Konno et al. demonstrated a non-linear inverse relationship between rotavirus infection and temperature in Japan. The relative importance of pathogens that thrive at lower temperatures appears to be greater in populations with higher standards of living, who have access to clean water and sanitation and for whom there is no clear and consistent evidence for peaks in all-cause diarrhoea in warmer months. This is in contrast to the situation of less well-off populations, where diarrhoea is usually more common in warmer, wetter months, as well demonstrated by clear summer peaks of diarrhoea in black, but not white, infants in Johannesburg during the 1970s.

For this assessment we defined developing countries as those with per capita incomes lower than the richer of the two study countries (Fiji) in the year 2000—approximately US$6000/year in 1990 US dollars. For such countries, we applied a dose–response relationship of 5% increase in diarrhoea incidence per degree centigrade increase in temperature, to both sexes and all age groups. This is consistent with the relationships derived from the two studies described above. We chose 5% rather than the arithmetic mean of the constants from the two studies (5.5%) for two reasons: (i) to avoid giving a false impression of precision based on only two estimates, each with their own confidence intervals, and (ii) in order to be conservative.

Although the confidence intervals around the estimates from the individual studies are relatively small, these clearly cover only a small range of climatic and socioeconomic environments. As described above, Checkley et al. showed that even in a single socioeconomic setting the temperature dependence of diarrhoeal disease may vary across the temperature range or, less plausibly, with non-climatic seasonal variations. This introduces uncertainty when extrapolating such a single relationship. There is also potential bias: if the temperature-responsiveness is indeed greater at low temperatures, extrapolation of an average value will tend to underestimate effects in areas that are on average colder, and overestimate in hotter regions. However, the average annual temperatures experienced by the residents of the areas to which we extrapolated this relationship was 20.3 °C, as calculated by averaging the temperatures in each 1°x1°grid-cell weighted by the population. This is in the mid-range of the temperatures experienced throughout the year in Lima (16-26°C), and cooler than those in the capital of Fiji (23-27°C). Our extrapolation tended towards being conservative, but with significant uncertainty. Therefore, we place a wide uncertainty range (0-10%) on this value.

As there are no such studies published for developed regions, we made the assumption that overall diarrhoea incidence in richer countries is insensitive to climate change, that is, 0% (—5 to +5%) change per degree centigrade temperature change. Relative risks for each country under each scenario were calculated by multiplying the projected increase in temperature by the relevant exposure-response value. The quoted estimate for each subregion is the population-weighted average of the relative risks for each country in the subregion. In addition to changes in baseline diarrhoeal disease over time, we assumed that the climate sensitivity of diarrhoea in developing countries will decrease as they become better off. For projections of relative risks for years after 2000, we used projections of future changes in GDP (WHO/EIP/GPE, unpublished data, 2000) to apply the relationship used above—that is, overall diarrhoea incidence does not respond to temperature in any country that attains a per capita GDP of at least US$ 6000/year. Relative risks for each time point were then calculated as above.

The quantitative estimates in this analysis were highly sensitive to the exposure-response relationship, around which there is substantial uncertainty due to the very small number of analysed time-series. The estimates could be rapidly improved by analysis of climate exposure-response relationships from sites from a wider climatic and socioeconomic spectrum. Future studies should also explicitly measure the degree to which economic development and improved levels of sanitation influence vulnerability to the effects of climate variation on diarrhoeal disease.

Malnutrition

Multiple biological and social factors affect the incidence of malnutrition, but one of the fundamental determinants is the availability of staple foods. Climate change may affect this through the balance of the (broadly negative) effects of changes in temperature and precipitation, and (broadly positive) effects of higher CO_2 levels on yields of major food crops. These effects are likely to vary markedly with geography: productivity is projected to increase in higher-latitude producers such as Canada and the United States, but to decrease closer to the equator. The global food trade system may be able to absorb these effects at the global level. However, climate change can be expected to have significant effects on food poverty in some regions, owing to variation in both productivity and in economic capacity to cope.

Crop models have been validated at 124 sites in 18 countries over a wide range of environments. However, estimation of changes in food availability (and by inference malnutrition) requires additional analyses of the geographical variation in effects on food production, and of food trade patterns. Only one modelling group has integrated a basic physiological model of climate change effects on region-specific crop production with food trade models, in order to make projections of the numbers of people actually at risk of hunger. All results presented here were based on the model described in Parry et al. and other work by the same group.

As for other potential effects of climate change, there is considerable uncertainty over the degree to which current relationships, such as those between climate and crops, and food trade systems will remain constant over time. The most important uncertainties probably relate to the ability of the world food trade system to adapt to changes in production. Although these are the most complete models currently available, they do not describe the likely effect of climate change on more complex pathways, such as animal husbandry, or the relative importance of fruit and vegetable production. These in turn may affect micronutrient deficiency.

Global distribution of temperature, rainfall and CO_2 were mapped for each of the alternative scenarios, as described above. Climate dose–response relationships have been defined for yields of major grain cereals and soybean, which account for 85% of world cereal exports.

Effects of temperature and precipitation, and the beneficial effects of higher CO_2 levels, have been defined using the IBSNAT-ICASA dynamic crop growth models. The exposure distributions described above were applied to these crop growth models, and the derived yield functions were extrapolated to other crops and regions on the basis of agro-climatic similarity.

These crop yield estimates are used as inputs for the Basic Linked System world food trade model. This consists of a linked series of 40 national and regional food models, representing food production, the effects of market forces and government policies on prices and trade, and trends in agricultural, economic and technological conditions over time.

The model is represented schematically and characteristics of this model are:

- no major changes in the political or economic context of world food trade or in food production technology;
- population growth to occur following the World Bank mid-range estimate i.e. 10.7 billion by the 2080s;
- GDP to accumulate as projected by EMF14 ;
- a 50% trade liberalization in agriculture is introduced gradually by 2020.

The model results in an estimation of national food availability. This is used to generate an estimate of per capita food availability in each country, assuming that this food is distributed among the population following a skewed (beta) distribution. The final model output is the number or proportion of the population in each subregion who do not have access to sufficient food to maintain a basal metabolic rate of 1.4, which is the Food and Agriculture Organization of the United Nations' (FAO) definition of undernourishment.

The model generates outputs for continents made up principally of vulnerable developing countries (thus excluding China and the former Soviet Union, countries in North America, and in western and eastern Europe). As these model continents do not map directly on to the sub-regions, we generated estimates for each subregion by calculating the proportion of the population that lives within each continent in the food availability model. Where more than 90% of the subregion population live within a single model continent, we quoted the model estimate for that continent. Otherwise an average was calculated, weighted by the distribution of the subregion population among the various continents. While the subregions mapped reasonably well on to the climate change/food availability model continents, the aggregation meant that some of the geographical variation in vulnerability was lost. Most notably, despite severe problems in some countries, EUR-C was assumed not to suffer from malnutrition as it lies within the "developed" European continent of the food availability model.

Preliminary analysis correlating estimates for the 1990s at the level of the model regions (data not shown) indicated that the model output was positively related to more direct measures of malnutrition, such as the incidence of underweight, and stunting and wasting in children aged <5 years, as measured by the WHO Global Database on Child Growth and Malnutrition. The aggregation of the food availability model means that this correlation was based on only a small number of independent data points. We therefore did not attempt to make any quantitative estimate of these relationships. Instead, the relative risk of the incidence of energy shortfall was interpreted as being directly proportional to the relative risk of suffering from the risk factor "underweight". The relative risk estimates were therefore applied to all diseases affected by the underweight risk factor. These include diarrhoea and malaria. We therefore assumed that these diseases are affected in two distinct ways by climate change—through meteorological effects on the pathogens and vectors, and through increased susceptibility of the human population due to undernutrition.

In common with most climate change impact assessments, the published studies do not quote uncertainties around the various relationships in the model, either separately or aggregated. Small variations in the initial conditions of a single climate model (the HadCM2 ensemble) generated only slight variations in projections of crop production and incidence of food shortfall. Using a different climate model (HadCM3), however, generated markedly different projections. These comparisons relate to only part of the possible error, as they do not address uncertainties in the trade and social components of the model.

From the published descriptions of the model, there is no reason to assume that the estimates generated were systematically biased either upward or downward. In the absence of formal sensitivity analyses of the complete model, however, uncertainty estimates are arbitrary. We presented the relative risks generated above as mid-range estimates, with the upper and lower range covering a complete adaptation to any changes in agricultural output (i.e. no change in risk), to a doubling of the estimate of the relative risk calculated above. However, this uncertainty range should be treated with caution. Priorities for the future are investigations of:

- variation in output when using a wider range of food production models as inputs to the food trade/availability models;
- sensitivity of estimates to the various climate scenarios;
- sensitivity analyses to estimate uncertainty around the exposure–response relationships;
- back-casting of climate/hunger relationships for verification of model accuracy;
- a finer geographical (e.g. country level) breakdown of the outputs of malnutrition models;
- correlations between model outputs and health outcomes at a high spatial resolution; and
- investigation of the synergistic effects of water availability and poverty on malnutrition.

Vector-borne Diseases

Viruses, bacteria, protozoa and helminths transmitted by biting insects and other intermediate hosts are among the most important causes of ill-health in tropical regions. The climate sensitivity of such diseases has long been recognized , and knowledge of the relationships has been used to help predict epidemics of vector-borne diseases since at least the early years of the last century. More recently, there have been many quantitative studies of the effects of climate variables on the population biology of vectors and pathogens in the laboratory, and on spatial and temporal variations in vector abundance and disease incidence in the field.

This climate sensitivity has prompted several studies correlating long-term changes in vector distribution or disease incidence with local climate trends, which apparently reflect global climate change. However, many of the inferences about resulting health consequences have been called into question, due to concurrent local changes in crucial non-climatic factors, such as human behaviour, disease reporting or control programmes.

As for other health effects, it is unlikely that simply correlating long-term trends in disease against trends in climate will soon (or perhaps ever) give unequivocal evidence of the effects of gradual climate change. As climate varies naturally between years, long time-series (i.e. two to three decades) will be needed for statistical tests of association in long-term trends. It is almost inevitable that non-climatic determinants of risk will also change over such long periods, either obscuring or offering alternative explanations for any effects of climate change. In addition, reliable multi-decadal time-series for vector-borne diseases in developing countries are rare, and it is problematic to interpret analyses of a few data sets as evidence of a general global pattern. Recent reviews therefore suggest that a general effect of climate change on vector-borne disease, while suspected and possible, is highly dependent on other non-climatic factors that act at a small scale.

Despite the practical problems in making direct correlations with recent trends, the extreme climate sensitivity of vector-borne diseases means that it is almost inevitable that they will respond in some way to climate change, in some settings and to some degree. The most reliable basis for estimating such changes should come from information on the relationships between variations in climate and disease in either the past or present. Several studies have used such data to model the effect of predicted climate change on either the distribution of vector-borne diseases, and/or measures of risk within existing or predicted newly endemic areas.

Whatever the quality of the data and modelling techniques used for predictions, they will remain contingent upon other determinants of disease. Socioeconomic conditions, control programmes, human immunity and the specific combinations of climate variables required by particular vector species or transmission cycles also affect disease incidence, and may be more important than global climate trends, particularly at small spatial scales.

The IPCC has reviewed the observed and predicted effects of climate variability and change in the context of the other factors listed above. On balance, the IPCC concludes that climate change is likely to expand the geographical distribution of several vector-borne diseases, including malaria, dengue and leishmaniasis to higher altitudes (high confidence) and higher latitudes with limited public health defences (medium/low confidence), and to extend the transmission seasons in some locations (medium/high confidence). For some vector-borne diseases in some locations, climate change may decrease transmission by reductions in rainfall or temperatures too high for transmission (medium/low confidence).

The associations between climate and infectious diseases have also been reviewed by the United States National Research Council. The review highlights the climate sensitivity of vector-borne diseases, and describes some studies modelling the potential consequences of future climate change on vector-borne diseases. The report re-emphasizes the need for caution when making future projections, and stresses the importance of public health provision in mitigating increases in incidence driven by climate change. No judgement is made on the likely effects of future climate change on specific diseases, owing to the limited evidence.

Aside from analyses of the potential effect of gradual changes in average climate conditions, predicted increases in climate variability, including possible increases in the frequency and intensity of El Niño events , may also affect vector-borne diseases. There is evidence that the El Niño cycle causes inter-annual variations in disease incidence in several areas. However, even in these sites, other determinants such as seasonal variations and control programmes exert a larger influence, and epidemic cycles also occur in areas where El Niño has little or no effect on climate, apparently driven by gradual post-epidemic waning of herd immunity. As the effect of climate variability differs so greatly between sites, global projections of the associations described so far are unlikely to be informative.

Modelling of Specific Diseases

Most modelling of the effects of climate change has focused on malaria, and to a lesser extent dengue. These are therefore the only vector-borne diseases considered here. Some preliminary modelling work has been carried out on schistosomiasis , but this is based on a relatively small data set and has not been validated against current distributions. Randolph and Rogers have also modelled the potential effects of climate change on tick-borne encephalitis (TBE) in Europe, demonstrating that increased temperatures are likely to reduce the endemic range. Although TBE is a relatively small public health problem in global terms, and was not covered in this assessment, this study does demonstrate that climate change may potentially decrease, rather than increase, the transmission of some diseases. Effects on other major vector-borne diseases have been investigated either qualitatively (e.g. Carcavallo and Curto de Casas 1996 for American trypanosomia-sis) or in terms of distribution of vectors rather than human disease.

Falciparum Malaria

Falciparum malaria is unusual in that several research groups have independently modelled the relationships between climate and disease distribution. The models used can be broadly classified as structural/ biological, based on aggregating the effect of climate on the individual components of the disease transmission cycle, or "statistical", derived from direct correlations between geographic or temporal variations in climate, and associated variations in disease incidence or distribution, either in the present or recent past.

Published biological models for the global distribution of falciparum malaria use laboratory data to define the relationship between temperature and the extrinsic incubation period of the parasite, and therefore the probability of completing development during the lifetime of the mosquito and completing the transmission cycle. Later models incorporate temperature effects on the survival probability and biting frequency of mosquitoes. In these later models, the various temperature-dependent relationships are aggregated into the entomological version of the equation for R_0, the number of cases arising from each new case in a completely susceptible population . Because of the lack of data on several key parameters, these are set as biologically plausible constants, allowing the calculation of the critical vector density required for sustainable disease transmission. This threshold is lower under more suitable climate conditions. The inverse of the critical density threshold, the "transmission potential", is used as a relative measure of transmission intensity under different climatic conditions. The models also assume a threshold level of transmission potential required to sustain transmission, which allows the identification of areas that are climatically suitable for transmission under both observed and projected climate scenarios.

These studies highlight the extreme climate sensitivity of several stages of the malaria transmission cycle. By aggregating these effects into a single measure related to R_0, they demonstrate that even small temperature increases could potentially cause large relative increases in risk, particularly at the edges of the distribution where temperature may be a limiting factor. They also suggest that those areas climatically suitable for *Plasmodium falciparum* transmission could expand substantially.

While valid for their original purpose as sensitivity analyses for relative changes in risk, these models are not ideal for defining the most probable changes in either geographical distribution or disease burden within endemic areas. Both outputs require the calculation of absolute rather than relative values of R_0, so as to identify areas where $R_0>1$, allowing disease transmission to persist. In these incomplete biological models, such calculations are partly dependent on parameter values that are arbitrarily defined in the absence of empirical data. As the models are based on temperature relationships derived from the laboratory, they also rely on the assumption that meteorological station data accurately represent the climatic conditions that mosquitoes and parasites experience in the field—which disregards the possibility that vectors might exploit microhabitats that are very different from those in meteorological stations. Since the outputs from these models have not been validated against current disease distributions, they were not used in this assessment.

In the absence of data to generate complete biological models of all stages in the transmission cycle, an alternative approach is to use statistical relationships to define only the distributional limits of disease. Although this approach does not allow disaggregation of the specific mechanisms driving the climate-sensitivity of vector-borne diseases, it is generally considered more objective than the use of incomplete biological models, in that model outputs are not dependent on arbitrarily defined parameter values.

The international MARA (Mapping malaria risk in Africa) collaboration generated a model that used a combination of biological and statistical approaches to define the limits of climate suitability for falciparum malaria in Africa. Laboratory data on the rate of development of falciparum parasites and laboratory and localized field observations of temperature effects on mosquito survival were used to define upper and lower thresholds for mean monthly temperatures, and winter minima, which would allow both mosquito survival and the completion of the parasite extrinsic incubation period during the lifetime of mosquitoes, thereby permitting transmission. Rainfall thresholds were defined by comparing regions with and without stable malaria transmission, which have similar temperature conditions but different precipitation profiles. In order to take account of uncertainty about the precise values of the upper and lower bounds of temperature and rainfall necessary for transmission, climatic conditions near to the thresholds were not defined as either entirely suitable or unsuitable. Instead, they were assigned a probability of suitability between 0 and 1, defined by a "fuzzy membership curve", which is assumed to follow a pre-specified sigmoidal shape between the plausible values for the upper and lower thresholds for each climate variable (e.g. a decreasing probability of suitability between mean temperatures of 32 and 40°C). These relationships were applied to high-resolution interpolated maps of climate throughout Africa to define areas that meet all suitability conditions (i.e. both temperature and rainfall) throughout the continent. For validation, model outputs were visually compared with independent high-resolution maps of the edges of the distribution, based either on field surveys or expert opinion. The model showed a good fit to the observed distributions in both southern Africa and East Africa.

The main advantages of this approach are that the model:

- describes only the cut-offs for any level of transmission, rather than quantitative estimates of transmission risk: it therefore does not rely on arbitrarily defined parameter values to complete the R_0 equation;
- represents uncertainty around the edges of the distribution;
- allows description of seasonal patterns of transmission, based on the suitability of individual months; and
- most importantly, has been compared with current and historical distribution maps which are apparently independent of the model building process.

The main caveats are:

- the reliance on laboratory data and a small number of field studies to define climate cut-offs;
- apparent subjectivity in at least one parameter estimate (the proportion of mosquitoes that need to survive the sporogonic cycle in order to maintain transmission, which defines the precise value of the lower temperature cut-off);
- the need to make an assumption about the shape of the "fuzzy membership curve"; and
- lack of systematic empirical validation (validation by visual comparison, rather than calculating diagnostic statistics).

While each of the assumptions can legitimately be questioned, the visual validation suggests that the data and assumptions used are at least reasonably accurate. Again, there are further caveats in using the model to try to describe the true global distribution of falciparum malaria (rather than just climatically suitable areas), either now or in the future. To use such models, it is necessary to make the assumptions that distributions vary directly with climate, without any interactive effect of control programmes, or socioeconomic conditions. The comparisons of model outputs with current data suggest that this is a reasonable assumption for most of sub-Saharan Africa, although control programmes have altered distributions in South Africa. The assumption is much less secure for other endemic regions, which are invariably richer.

The relationship between climate variables and the global distribution of malaria can also be defined in statistical terms. Rogers and Randolph converted WHO maps of the limits of reported malaria distribution in the 1990s into 0.5°¥0.5° resolution grids, coding each cell as either endemic or non-endemic. These grid maps were overlaid on 0.5° grid maps describing climate surfaces for the period 1961–1990. A statistical model was generated by randomly selecting a subsample of 50% of the observations of disease presence or absence. First, grid-cells were assigned by applying *k*-means clustering to six groups based on climatic similarity, thereby allowing for potentially different climate–disease relationships in different ecological zones. Stepwise discriminant analysis was applied to find the combination of temperature, rainfall and humidity parameters that gave the greatest statistical differentiation between positive vs negative grid cells within each cluster. The model was then applied to the remaining 50% of the observations to assess model accuracy. The fit of the predicted distributions to the observed WHO malaria maps was found to be significantly better than for previous falciparum malaria models: 77.71% of grid cells correctly predicted for this model vs 75.79% for Martin and Lefebvre and 67.26% for Martens et al..

The principal advantage of this approach is that it is entirely data driven. The relationships between global climate and malaria distributions are defined using transparent statistical techniques, applied to both disease and climate data from throughout the globe. They therefore do not require either unsupported assumptions of parameter values in order to complete a biological model, or global extrapolation of data from a limited number of observations. In addition, by mapping the observed distribution rather than climatically suitable areas, such analyses do not require the assumption that distributions are defined only by climate.

Despite these advantages, the quality of the data available at the global scale places several limitations on global statistical models. WHO maps of observed distributions are based on a combination of field observations and expert opinion to draw "inclusive" boundaries of the extremes of distributions. In many cases the maps define large areas as endemic (e.g. all of sub-Saharan Africa north of South Africa, Namibia and Botswana), although significant areas are actually disease free, often apparently due to unsuitable climate. WHO maps (and therefore the statistical relationships) also do not differentiate between malaria caused by *P. falciparum* and *Plasmodium vivax*, which have quite different sensitivities to temperature. It is unclear what effect these two simplifications may have on the climate sensitivity of the model.

For this assessment, projected changes in temperature and rainfall under each of the alternative climate scenarios relative to the baseline climate were mapped at the resolution of the HadCM2 climate model (3.75° longitude by 2.5 ° latitude). Maps of future climate were then generated by adding these values to maps of baseline climate for the 1961–1990 climate at 0.5° resolution.

An adapted version of the MARA climate model described in detail above was used to generate the mid-range estimates for this assessment. This decision was based on the independent (though continental rather than global) verification, plus the fact that the model was developed and tested using data from throughout Africa (where the overwhelming majority of the burden of malaria currently occurs, and where accuracy is therefore most important), and the potential for developing projections of increased force of infection and disease incidence in the near future.

The model was applied to the global climate maps for baseline climate, and for the unmitigated emissions scenario for the 2020s and 2050s. Relative risk estimates presented here were the ratios of the projected population at risk in each subregion under climate change, relative to the population at risk under the 1961–1990 climate. As an approximation, estimates for the s750 and s550 scenarios were derived proportionately by multiplying the relative risks under unmitigated climate change by the ratio of global temperature change under each scenario/change under unmitigated emissions. This model gave considerably larger estimates of changes in population at risk than the statistical model of Rogers and Randolph , which predicted approximately no overall change under an unmitigated emissions scenario by the 2050s. In the absence of further comparisons and formal uncertainty assessments, our lower range estimate therefore included the possibility of no change in risk in any subregion. The upper range estimate is a doubling of the mid-range estimate from the MARA model. We emphasize that given the difficulties in validating any specific model and its suitability for extrapolation to other subregions, the choice of MARA climate model over the other possible models was somewhat arbitrary.

In addition, the calculation of disease burdens requires estimates of change in incidence within each subregion, rather than population at risk. In the absence of models for changes in malaria incidence within endemic regions, we therefore made the assumption that relative changes in incidence will vary in direct relation to predicted changes in population at risk—that is, a doubling of the population at risk within a region will lead to a doubling of the clinical disease incidence.

These measures are related in broad terms: countries or regions with higher populations at risk tend to have higher incidence and disease burdens. However, this relationship is a crude generalization, as it assumes that these relationships will remain constant as the population at risk expands or contracts. This may lead to underestimation of effects, if there is an increase in transmission within already at-risk populations, driving up infection incidence. Alternatively, this relationship may overestimate risk, depending on the extent to which increasing vectorial capacity promotes herd immunity , and causes first infections to occur earlier in life, when patients suffer less severe clinical symptoms for some diseases, potentially conferring immunity on the more clinically vulnerable older age-groups. In addition, socioeconomic conditions and control programmes clearly influence vector-borne diseases.

By applying relative risks to baseline incidences of zero in some sub-regions, our assessment did not allow for the spread of disease from endemic subregions to non-endemic subregions. This is a reasonable, but conservative assumption: non-endemic subregions have better developed health systems and a less amenable socioeconomic environment, in addition to usually being cooler. These factors may protect against re-establishment of vector-borne disease transmission , providing they are maintained.

Dengue

Several studies have explored the relationship between climate and the distribution of intensity of dengue transmission. Most are derived from a series of biological models that relate climate variables to determinants of the population biology of *Aedes* vectors and dengue transmission. Adaptations of these models to map climate relationships at the global scale have used field and laboratory data to define the relationships between temperature and the length of the gonotrophic cycle (and the associated feeding frequency), larval weight (and therefore the need to take multiple feeds within a single cycle), and the extrinsic incubation period of dengue virus within the vector. Mosquito survival, human biting habit and the duration of human infectiousness are set as temperature-independent constants, with parameter values defined using field data from a number of sites.

As for the biological models of malaria, these relationships are aggregated into a simplified version of the R_0 equation, excluding measures of vector abundance. This equation is again used to define the critical density threshold, and its inverse, "transmission potential" or "transmission intensity". The model has been applied to local climate data in a series of sites, and the rank order of monthly predicted values of transmission potential showed good correspondence with the observed seasonal distribution of dengue cases.

These models have similar characteristics to the biological models for malaria. By aggregating temperature effects from various stages of the transmission cycle, they demonstrate that overall dengue risk is likely to be extremely sensitive to even small changes in temperature. Sensitivity analyses show that a 2°C increase in global temperature would lead to large (commonly 1-5 times) increases in transmission intensity in many regions of the world. However, these models are also subject to the same caveats. As there are no data available on mosquito abundance throughout the globe, or the relationship between vector abundance and climate variables, the transmission potential remains a relative rather than absolute measure of R_0. It is consequently unsuitable for defining the conditions under which transmission can persist, and therefore mapping the distributional limits of dengue either now or in the future. In addition, although the validation exercise demonstrates that transmission potential is correlated to the incidence of infection and clinical disease, in that months with higher transmission potential have more cases, the quantitative relationships have not been explicitly defined. It is therefore difficult to interpret exactly what effect a doubling of transmission potential (for example) would have on the burden of disease caused by dengue.

Methods similar to those used by Rogers and Randolph for malaria have recently been applied to define the relationship between multiple climate parameters and the reported geographic distribution of dengue. Hales et al. used WHO data to map reports of dengue transmission during the period 1975–1996 at the level of the country, or smaller administrative area when this was specified in the report. These were converted into grid maps at 0.5° resolution. Logistic regression analysis was then used to correlate presence or absence of reported transmission against the average values of various climate parameters (monthly average rainfall, vapour pressure, and maximum, minimum and mean temperature) for the corresponding grid-cells throughout 1961–1990. Vapour pressure (approximately equivalent to absolute humidity, and reflecting both temperature and precipitation) gave the best discrimination between areas with and without reported dengue transmission, with no other climate variables adding significant explanatory power. The accuracy of the model was assessed by cross-validation, repeatedly using 95% of the data to generate predictions for the remaining 5%. The model gave correct predictions of observed presence or absence for an average of 89–92% of grid-cells, with the precise accuracy depending on the radius over which vapour pressure was considered. The limitations of this model are similar to the statistical model for malaria. To be consistent with the choice and exclusion of models for malaria, the model of Hales et al. , which is similar to the excluded model of Rogers and Randolph , was not used in the estimates of disease burden.

References

Kalkstein LS, Smoyer KE (1993) The impact of climate change on human health: some international implications. *Experientia*, 49:969–979.

Konno T, Suzuki H, Katsushima N et al. (1983) Influence of temperature and relative humidity on human rotavirus infection in Japan. *Journal of Infectious Diseases*, 147:125–128.

McMichael AJ, Kovats S, Martens P et al. (2000) *Climate change and human health: final report to the Department of Environment,Transport and the Regions*. London School of Hygiene and Tropical Medicine/ICIS, London/Maastricht.

Rogers DJ, Randolph SE (2000) The global spread of malaria in a future, warmer world. *Science*, 289:1763–1766.

Rosenzweig C, Parry ML (1994) Potential impact of climate-change on world foodsupply.*Nature*, 367:133–138.

Bibliography

Angell, J.K., (1988). Variations and trends in tropospheric and stratospheric global temperatures, 1958-87. *J. Climate*, 1, pp. 1296-1313.

Barrett, S. and Graddy K. (2000) Freedom, growth and the environment. *Environment and Development Economics* 5(4):433–456.

Barry, R.G. & Chorley, R.J., (1992). *Atmosphere, Weather & Climate*, 6th edn. Routledge, London. 392pp.

Bell, M. M. (2004) *An Invitation to Environmental Sociology,* 2nd edn. Thousand Oaks: Pine Forge Press.

Benton, T. (1998) Sustainable development and the accumulation of capital: reconciling the irreconcilable? In Dobson A. (ed.) *Fairness and Futurity*. Oxford: Oxford University Press.

Bradley, R.S. & Jones, P.D., (1992). *Climate Since A.D. 1500*. Routledge, London, 679pp.

Brown, L. (2004) *Outgrowing the Earth*. New York: W. W. Norton

Brunkhorst, D., Bridgewater, P. and Parker, P. (1997) The UNESCO biosphere reserve program comes of age: learning by doing; landscape models for sustainable conservation and resource use. In Hale P. and Lamb D. (eds) *Conservation Outside Nature Areas*. University of Queensland, pp. 176–182.

Carney, D. (1998) *Sustainable Rural Livelihoods.* London: Department For International Development

Kalkstein LS, Smoyer KE (1993) The impact of climate change on human health: some international implications. *Experientia*, 49:969–979.

Kemp, D.D., (1994). *Global environmental issues: a climatological approach*, 2nd ed. Routledge, London, 224pp.

Konno T, Suzuki H, Katsushima N et al. (1983) Influence of temperature and relative humidity on human rotavirus infection in Japan. *Journal of Infectious Diseases*, 147:125–128.

Lamb, H.H., (1982). *Climate, History and the Modern World.* Methuen & Co., Ltd., London.

Linacre, E., (1992). *Cliamte data and resources: a reference and guide.* Rouledge, London, 366pp.

McMichael AJ, Kovats S, Martens P et al. (2000) *Climate change and human health: final report to the Department of Environment,Transport and the Regions.* London School of Hygiene and Tropical Medicine/ICIS, London/Maastricht.

Midlarsky, M. (1998) Democracy and the environment: an empirical assessment. *Journal of Peace Research* 35(3): 341–361.

Mitchell, J.M. *et al.*, (1966), *Climate Change. Technical Note No. 79.* World Meteorological Organisation.

Mol, A.P.J. (2001) *Globalization and Environmental Reform: The Ecological Modernization of the Global Economy.* Boston: MIT Press.

Parry, M.L., (1990). *Climate Change & World Agriculture.* London: Earthscan Publications Ltd.

Pickering, K.T. & Owen, L.A., (1994). *An introduction to global environmental issues.* Routledge, London, 390pp.

Rogers DJ, Randolph SE (2000) The global spread of malaria in a future, warmer world. *Science*, 289:1763–1766.

Rosenzweig C, Parry ML (1994) Potential impact of climate-change on world foodsupply.*Nature*, 367:133–138.

Schnaiberg, A. (1980) *The Environment, From Surplus Scarcity.* New York: Oxford University Press.

Trewartha, G.T. & Horn, L.H., (1980). *An Introduction to Climate.* New York: McGraw-Hill.

Turner, R.K., (1995) *Sustainable Development and Climate Change*, Working Paper PA 95-01, Centre for Social & Economic Research on the Global Environment (CSERGE).

United Nations Environmental Programme / World Meteorological Organisation (UNEP/WMO), (1992). *Framework Convention on Climate Change.* United Nations, Geneva.

Uphoff, N. (ed.) (2002) *Agroecological Innovations.* London: Earthscan

Ward, H. (2006) International linkages and environmental sustainability: the effectiveness of the regime and IGO networks. *Journal of Peace Research* 43:149–166

Wigley, T.M.L. & Barnett, T.P., (1990). Detection of the Greenhouse Effect in the Observations. In: *Climate Change: The IPCC Scientific Assessment,* Intergovernmental Panel on Climate Change (IPCC), 1990a. Houghton, J.T., Jenkins, G.J. & Ephraums, J.J. (eds.). Cambridge University Press, Cambridge, pp. 239-255.